CONTENTS

EXECUTIVE SUMMARY | 5

1 INTRODUCTION | 7

2 INVENTORY ANALYSIS OF MICRO-WIND TURBINE SYSTEMS | 9
  2.1 Introduction | 9
  2.2 University of Bath LCA data | 9
  2.3 System boundaries | 9
  2.4 Recycling | 9
  2.5 Results | 10
  2.6 Comparison with LCA data for other turbines | 10
  2.7 Installation, maintenance and operation of the micro-wind systems | 12

3 ESTIMATION OF TYPICAL URBAN WIND RESOURCE | 14
  3.1 Introduction | 14
  3.2 Wind resource – adjustment factors for urban environments | 18

4 ELECTRICITY GENERATION BY BUILDING-MOUNTED WIND TURBINES IN TYPICAL URBAN SCENARIOS | 28
  4.1 Introduction | 28
  4.2 Methodology for the electricity calculation | 28
  4.3 Results | 29
  4.4 Conclusions | 32

5 CO₂ PAYBACK FOR DOMESTIC MICRO-WIND TURBINES IN URBAN ENVIRONMENTS | 33

6 LIFE CYCLE COSTS AND FINANCIAL PAYBACK FOR MICRO-WIND TURBINES | 37
  6.1 Introduction to life cycle costing | 37
  6.2 What costs are taken into account when undertaking LCC for a wind turbine? | 39

7 DISCUSSION AND CONCLUSIONS | 42

8 FURTHER WORK | 44

9 REFERENCES | 45

APPENDIX A - UK Mean annual wind speed at 25 m above ground level | 46
APPENDIX B - Locations for which the BREVe scaling factors were obtained | 47

MICRO-WIND TURBINES IN URBAN ENVIRONMENTS

An assessment

Richard Phillips, Paul Blackmore, Jane Anderson,
Michael Clift, Antonio Aguiló-Rullán and Steve Pester

IHS **bre** press

bretrust

The mission of BRE Trust is 'Through education and research to promote and support excellence and innovation in the built environment for the benefit of all'. Through its research programmes the Trust aims to achieve:
- A higher quality built environment
- Built facilities that offer improve functionality and value for money
- A more efficient and sustainable construction sector, with
- A higher level of innovative practice.

A further aim of the Trust is to stimulate debate on challenges and opportunities in the built environment.

BRE Trust
Garston, Watford WD25 9XX
Tel: 01923 664000
secretary@bretrust.co.uk
www.bretrust.co.uk

BRE Trust and BRE publications are available from
www.brepress.com
or
IHS BRE Press
Willoughby Road
Bracknell RG12 8FB
Tel: 01344 328038
Fax: 01344 328005
brepress@ihs.com

Published by IHS BRE Press for BRE Trust

FB17
© Copyright BRE Trust 2007
First published 2007
ISBN 978-1-84806-021-0

EXECUTIVE SUMMARY

The threat from climate change coupled with concerns over energy security are driving growth in locally based small-scale electricity generation (micro-generation), especially those based on renewable energy sources. UK Government is backing this drive through a combination of regulation, grants and targets for reducing CO_2 emissions, including a challenge to the construction industry to deliver 'zero carbon' homes by 2016.

One of the technologies being considered is micro-wind turbines designed to be mounted on new and existing homes for small-scale electricity generation. However, there is little experience of the operation of such turbines mounted on domestic buildings in urban environments and hence little objective data about their actual performance in terms of power generation, service life and maintenance requirements. This has led to concerns that, in some environments, the installation of micro-wind turbines on housing could increase carbon emissions rather than reduce them.

A number of studies [1, 2] are currently being established to monitor the output from actual installations, but it is considered that it will be necessary to collect data for at least a year to give a meaningful indication of the annual output and for somewhat longer periods to determine the true effectiveness of these turbines. This study is intended to give a more rapid indication of the likely effectiveness of such turbines, pending this empirical data, by comparing estimates of the environmental impact of their manufacture, installation and maintenance with the impact saved by the electricity that could expected to be generated during their useful life, in some representative urban situations.

Clearly, the output from wind turbines is highly dependent on the local wind conditions. For micro-wind turbine systems, the assessment of the wind resource is normally based on wind speed databases, which do not take account of the surface roughness effects of the buildings in an urban environment or the local effects around the building on which the turbine is mounted. The use of such data can give an over-optimistic estimate of the wind resource. This study investigates a novel method for estimating the wind resource in urban areas and uses this, along with data from a wind tunnel study of the expected wind speeds around typical house roofs, to estimate the electrical output that could be expected from some typical micro-wind turbines in three different types of urban area within the UK (Manchester, Portsmouth and Wick).

The study has highlighted that, in addition to the initial embodied carbon and efficiency of the turbine, the payback period is highly sensitive to relatively small changes in one or more of a large number of variable factors, in particular:
- the local wind conditions
- the size of conurbation and the position within the urban terrain
- the type of building on which the turbine is mounted and the mounting position
- the proximity of the surrounding buildings
- the transport associated with installation and maintenance
- the maintenance regime
- the expected service life of the turbine.

With a study of this type, it is vital to recognise that there are many caveats to the findings but the study has indicated that:
- There is a large variation between the likely output of a micro-wind turbine in a large city such as Manchester (less than 150 kWh per year) and a windy location such as Wick in northern Scotland (up to about 3000 kWh per year).
- There is a large variation between the outskirts and town centres in windy locations such as Portsmouth and Wick.

- In suitable windy locations, micro-wind turbines can generate sufficient energy to pay back their carbon emissions within a few months to a few years (depending on the efficiency, lifetime and maintenance regime).
- In large urban areas such as Manchester, even with very favourable assumptions about efficiency, lifetime and maintenance, micro-wind turbines may never pay back their carbon emissions.

A brief life cycle cost analysis of the systems included in this study suggests that, at today's prices, financial payback is unlikely for all but the most durable, efficient and low maintenance turbines, even in favourable urban locations.

This work confirms the need for a more rigorous method for estimating the electricity generated from building-mounted micro-wind turbines and for research and innovation in the technologies, including:

- improved methods for estimating the wind resource in urban environments and close to building roofs. The methods proposed in this report should provide a suitable basis for these.
- a method to account for the effects of wind turbulence on the power generated by micro-wind turbines
- performance optimisation
- durability
- improved means of access for maintenance and repair.

Research and guidance will also be needed on planning and urban design to maximise the effectiveness of the turbine installations, while minimising the visual and noise impacts, and any structural implications for the building on which the turbine is mounted.

1   INTRODUCTION

Part of the Government's energy policy is to increase the contribution of electricity supplied by renewable energy to 10% by 2010. Initiatives intended to help achieve this target include the promotion of locally based small-scale electricity generation (so called micro-generation), such as solar photovoltaic (PV) cells and small wind turbines. Government-sponsored programmes, such as the Low Carbon Building Programme[1], offer grants for the installation of micro-generation systems, including small wind turbines. These grant programmes, along with media interest in a new generation of micro-wind turbines specifically designed to be attached to buildings, are leading to an increase in the installation of domestic wind turbines.

It is clearly important to ensure that the measures being promoted will generate sufficient energy during their service life to compensate for the environmental impact related to their manufacture, installation and maintenance, and hence that they will be beneficial in reducing greenhouse gas emissions.

A number of studies and reviews have been conducted into to the effectiveness of wind generation, in terms of energy balance, and also its ability to help meet the UK's drive to reduce greenhouse gas emissions. However, these studies have focussed on larger scale wind generation schemes, and are not specifically relevant to 'micro-wind' generation installations, in particular to installations on typical houses, which are generally rated at 1.5 kW or less.

Clearly, the output from wind turbines is dependant on the local wind conditions. The power in the wind is due to its kinetic energy and is proportional to the cube of the wind speed. Hence, relatively small reductions in wind speed greatly reduce the available power (e.g. halving the wind speed reduces the available power eight times). This is especially significant for small-scale domestic installations in built-up environments where surrounding buildings can dramatically reduce the prevailing wind speeds. The presence of other buildings close to a micro-wind turbine will also increase the turbulence of the wind, which will further reduce the turbine output.

For large-scale wind turbine installations extensive wind monitoring is conducted before sites are selected. Such assessments are rarely possible for small urban installations and predictions are often based on wind speed data, such as that from the Department for Business, Enterprise & Regulatory Reform (BERR) (formerly Department of Trade and Industry) wind speed database, commonly known as the NOABL database. This gives estimates of the annual mean wind speed throughout the UK on a 1 km grid. However, the data is the result of an air flow model that estimates the effect of the gross topography on wind speed but does not account for effects such as the surface roughness caused by the buildings in an urban environment or the local effects around individual buildings. As a result estimates based on this data tend to give very optimistic results in urban locations.

To investigate the effectiveness of building-mounted micro-wind turbines in urban environments the performance of some typical installations of this type has been modelled in Manchester, Portsmouth and Wick. Manchester was chosen as a large inland city in an area of relatively low mean wind speeds (see map in Appendix A) and is considered representative of other such conurbations (e.g. Birmingham, Nottingham and to some extent London). Portsmouth was chosen to represent medium sized conurbations in coastal areas

---

[1] The Department for Business, Enterprise and Regulatory Reform (BERR) has established a third party certification scheme for microgeneration technologies and installers. This scheme requires installers to sign up to an Office of Fair Trading code of conduct and to provide homeowners with information on the potential energy generation but does not currently consider the CO_2 payback. Details of the scheme can be found at www.ukmicrogeneration.org.

and areas with moderate mean wind speeds and Wick was chosen to represent small towns in areas with high mean wind speeds. London was considered atypical due to its large size but it is anticipated that the attenuation of wind speed due to surface roughness found for Manchester would be exacerbated for London and hence the wind resource would be even less.

This study includes:

- estimates of the environmental impacts, including embodied energy and consequential greenhouse gas emissions (as equivalent carbon dioxide (CO_2) emissions), related to the construction, installation, maintenance and disposal of some typical micro-wind turbine systems
- the development of a method for estimating the wind resource in urban environments, which has been used to estimate the wind resource at locations in and around three towns (Manchester, Wick and Portsmouth)
- a wind tunnel study, reported separately[3], to investigate the effect of the house on which the turbine is mounted, and any surrounding buildings, on the wind speed at the turbine
- estimates of the expected annual electricity generation for some typical installations of these turbines, taking account of the power curves for the turbines, the wind resource, and building/mounting effects.

These have been used to investigate the likely financial and CO_2 payback periods required to balance the environmental impact of these typical turbine installations, over their anticipated service lives.

2 INVENTORY ANALYSIS OF MICRO-WIND TURBINE SYSTEMS

2.1 INTRODUCTION

The Department of Mechanical Engineering at the University of Bath is currently conducting a life cycle assessment (LCA) of some small wind turbines as part of a 4-year Engineering and Physical Sciences Research Council (EPSRC) funded project on Sustainable Power Generation and Supply (SUPERGEN Project). The LCA covers both the turbine and the electrical equipment necessary to connect to the national grid.

At this time, the University of Bath has received data (provided confidentially) for only one micro-wind turbine (referred to in this report as Micro-wind system 1). BRE has reviewed the Life Cycle Assessment carried out by Bath and adapted it where relevant to meet the 2007 BRE Environmental Profiles Methodology [4]. Additionally, BRE has compared the results with a published LCA [5] for another system (referred to here as Micro-wind system 3).

2.2 UNIVERSITY OF BATH LCA DATA

The LCA data from the University of Bath has been reviewed to ensure that it covers the relevant life cycle stages and system boundary required and represents a full picture of the embodied impact of the turbine. The methodology used and the data chosen have been reviewed to ensure that they are representative and defendable.

2.3 SYSTEM BOUNDARIES

The Bath study has used data for the input materials, including all the environmental impacts associated with extraction, processing and manufacture to delivery of the wind turbine to the customer. In the Bath study, upstream data (the LCA data for input materials) has been sourced mainly from the Swiss Government-funded ecoinvent database [6]. This is a well respected source of LCA data which BRE use for much of their LCA studies. However, BRE hold data from UK trade associations for most UK construction materials that is recent and specific to UK consumption. In this review, BRE have linked to this data where relevant.

Data for the material components and composition of the Micro-wind system 1 have been obtained from the manufacturer and augmented, where required, with generic industry data. Data was obtained by visiting the turbine factory and examining the technical drawings for the turbine and also by weighing the turbine components. In the absence of other data, data for an inverter owned by the University of Bath was used, and the constituent materials were estimated by electrical engineers at the university. Where a breakdown of the components in a part which had been imported from elsewhere was not available, an estimate was made using previously published LCAs, data held within the SimaPro LCA software [7] and other available industry data.

Energy and water use in the factory has not been included because it was not available. As only the assembly takes place in the factory, heating and lighting only need to be considered. While it is probable that this would have little overall effect on the turbine impact it would be desirable to collect this data in the future for the sake of completeness and certainty.

2.4 RECYCLING

Metal products can be manufactured from primary sources (e.g. mineral ores) or from recycled material. Using recycled material has a proven benefit in reducing overall environmental impact. However, because of the high value of scrap metal, there are typically very high recycling rates, of the order of 95%, and BRE believe that metal recycling is currently optimised, in the sense that sourcing a 100% recycled metal will not increase the amount of recycling globally, but just displace it. In this study therefore, the data has been reviewed by BRE so that metals have been modelled based on the typically used percentage of recycled content for each product type – for example, copper wire is manufactured exclusively from primary copper, whereas other copper products may have significant recycled input.

Micro-wind system 1 contains a considerable mass of metals including aluminium, steel, stainless steel and copper. These materials are likely to be recycled when they enter the waste stream at the end of life of the turbine. There are a number of approaches for modelling the benefits of this recycling within LCA studies, for example:

- **Value allocation**
  A proportion of the impact of primary manufacture can be passed on to any recycled metal, based on the relative values of the primary and secondary metals.

- **System expansion**
  Any recycling can be viewed as offsetting primary manufacture, so that a deduction is made that is equivalent to the impact of primary manufacture.

- **Multicycling approach**
  As metals are extensively recycled, the impact of primary and recycled manufacturing is averaged over a large number of cycles to give an 'average' manufacturing impact for both primary and recycled metals.

All of these methods are recommended in ISO 14044:2006 Environmental management – Life cycle assessment – Requirements and guidelines [8], but BRE favour the use of value allocation. Within the 2007 Environmental Profiles Methodology [4], BRE have modelled the allocation of primary manufacturing to future recycling, based on value, and this is included in the BRE Environmental Profiles Database [9].

2.5   RESULTS

The impact of manufacturing the Micro-wind system 1, including fixings and inverter, is shown in Figure 1. It is calculated in Ecopoints, a weighted measure of environmental impact.

From this, it can be seen that the largest impacts are associated with the inverter, the attachment to the building (mainly the mast), and the hub assembly. The actual turbine blades (made of glassfibre-reinforced polyamide) have only a very small impact. The impact of the inverter is mainly due to the significant mass of the stainless steel cover although the greatest mass of material in the inverter has been assumed to be cast iron. For the attachment to the building, it is the mast, made of galvanised steel, which causes most of the impact.

The overall system weighs approximately 45 kg, just over half of which is steel, with significant amounts of other metals. Included is the inverter weighing 16 kg based on data for a commonly used inverter for these types of system with a stainless steel cover.

In terms of climate change impacts, Micro-wind system 1 is responsible for 180 kg CO_2 equivalent (100 years) during its manufacture.

In terms of environmental impact, the system is responsible for 1.3 Ecopoints during its manufacture.

2.6   COMPARISON WITH LCA DATA FOR OTHER TURBINES

The data for Micro-wind system 1 has been reviewed to understand whether the turbine modelled is representative of micro-wind installations in the UK. This has been done by comparing it with publicly available data on the materials used in a turbine of a similar specification (Micro-wind system 2) and a published LCA study for another higher rated micro-wind turbine (Micro-wind system 3).

Micro-wind system 2

This system has a similar diameter to Micro-wind system 1, and manufacturer's data states a mass for the turbine assembly of 25–35 kg, with an aluminium generator and epoxy coated casing, and a galvanised steel mast of variable height. Blades and nose cone are of glassfibre-reinforced polyamide; the hub and boss are of anodised aluminium; the tail fin assembly is a steel box section; and the tail fin is an aluminium plate. The inverter for this system has a mass of 25–40 kg depending on the model. These results suggest a considerably heavier system than Micro-wind system 1 (45 kg total), of between 50 and 75 kg without

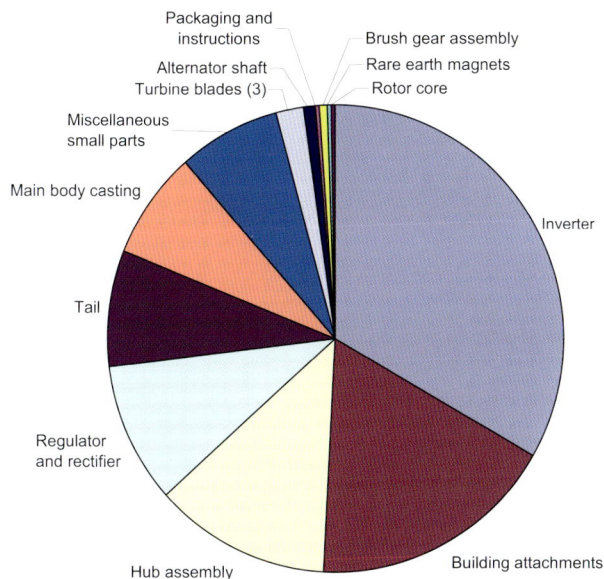

Figure 1. Environmental impact of Micro-wind system 1.

the mast and fixings. These results suggest a considerably heavier system than Micro-wind system 1 (45 kg total), of between 50 and 75 kg without the mast and fixings, which would need to be heavier to support the heavier turbine. For the purposes of the study, the data for Micro-wind system 1 has been adapted to model the manufacturing impacts of Micro-wind system 2 using factors based on the increased mass of the turbine assembly and inverter and building attachments. These are likely to produce a best case scenario as the factors are based on lower mass estimates, but without detailed figures we are unable to model the system accurately.

In terms of climate change impacts, Micro-wind system 2 is responsible for 299 kg CO_2 equivalent (100 years) during its manufacture. In terms of environmental impact, Micro-wind system 2 is responsible for 2.2 Ecopoints during its manufacture.

Micro-wind system 3

The third micro-wind system studied was modelled by the University of Edinburgh. BRE have not had access to this study, other than as published [5], so the data and methodology behind the study may not be directly comparable with that in the first study. This study provides some indicative LCA data for the raw materials that have been used, but they appear to have omitted data for processing and fabrication. This turbine uses a carbon fibre reinforced epoxy resin which appears to have more impact than the glassfibre-reinforced polyamide turbine used in Micro-wind system 1. This study includes assumptions for the installation, operation and maintenance phases of the life cycle, although they are small in comparison with the impact of the turbine components.

The total mass of the system is not given in the report, but manufacturer's literature suggests the mass of the turbine is 50 kg and the mass of the mounting system 40 kg. The mass of the inverter which has been included within the assessment is not given but has been assumed to be 40 kg. This gives a total mass of 130 kg, nearly three times the mass of Micro-wind system 1, although the expected output is about 1.7 times more and the diameter of the blades about 1.2 times more.

The data in the published report indicate that the primary energy for the life cycle of the system (including installation) is 22.8 GJ excluding recycling or 15.5 GJ including the benefit of recycling, and that CO_2 emissions for the life cycle of the system (excluding recycling) is 2428 kg. It should be noted that these figures assume that the aluminium used is primary aluminium (BRE normally assume that a typical mix of primary and secondary aluminium is used within construction components, which would reduce this impact). However, it should also be noted that the figure for CO_2 emissions is for CO_2 only and not for total greenhouse gas emissions.

The report also indicates that these figures include allowances for installation, operation and maintenance of the turbine. To allow correct comparison with the other systems in this report these allowances have been deducted and the impact of recycling has been included. This results in an impact of 1444 kg CO_2.

2.7   INSTALLATION, MAINTENANCE AND OPERATION OF THE MICRO-WIND SYSTEMS

In addition to the impact of manufacturing a turbine, there are also impacts associated with transporting it to the site and installing, maintaining and disposing of it at the end of its life. Several scenarios have been used to model these different aspects, which have been assumed to be the same for all turbine types.

For delivery of the turbine, two scenarios for the distance travelled by the turbine assembly from manufacturer to site have been used – a short distance of 100 km and a long distance of 400 km. In both instances, it has been assumed that the turbine is delivered by courier, so only a proportion (10%) of this distance would be allocated to the turbine. Delivery by van has been assumed.

Little information was available from the manufacturers about the inspection and maintenance required for the turbine systems considered, and the claimed service lives varied between 10 and 20 years (see Section 6). Turbines need to be mounted as high as practical above the building, so they are generally mounted on fixed masts attached to a high point of the building. In view of the weight of these units, the required mounting positions and the types of mounting arrangement provided, access for installation and maintenance by ladder would normally be impractical and contrary to health and safety requirements. It is therefore assumed that a hydraulic platform (cherry picker) or scaffolding tower is used.

In the absence of specific advice from manufacturers on the frequency of maintenance, we have assumed that a maintenance visit, requiring access to the unit, at least once every three years is needed to maximise the service life of the unit and to allow the necessary safety inspection of the blades, mountings and bearings etc.

For the purposes of this study, three-yearly maintenance is assumed to result in a service life of 20 years for each of the systems. Without maintenance, the service life is assumed to be 10 years. To investigate the impact of this, three options have been studied:
- 1: Installation and maintenance using a cherry picker
- 2: Installation and maintenance using a scaffolding tower
- 3: Installation with no routine maintenance.

In environmental terms, each installation or maintenance visit is assumed to consist of two separate journeys – the first of the equipment (cherry picker or scaffold tower) and the second of the maintenance engineers. No allowance has been made for parts that may need to be replaced at maintenance visits, as no advice on replaceable parts was available from the manufacturers. However, it is considered that the environmental impact of such parts is likely to be small in comparison with the impact of the maintenance visit itself.

Again, two scenarios have been modelled, with short and long travel distances as described below:
- We have assumed that the engineers travel round-trip distances of 50 km (short) or 200 km (long) by car.
- We have assumed that the cherry picker is roof mounted on a 3.5 t van and makes a round-trip of 25 km (short) or 100 km (long) Operation of the cherry picker has been ignored due to lack of data and the assumption that it has low impact compared to the transport. The size of the cherry picker (7.5 or 13 m) has been assumed not to have any significant impact on the result.
- For the scaffolding tower, we have assumed it travels 15 or 50 km round trip in a 3.5 t van.

These scenarios and options have been combined to produce the following options, which have been modelled:
- **Option 1 short**: Delivery, installation and 3-yearly maintenance by cherry picker with short distances for all transport. 20 year life for turbine (20 years output)
- **Option 1 long**: Delivery, installation and 3-yearly maintenance by cherry picker with long distances for all transport. 20 year life for turbine (20 years output)
- **Option 2 short**: Delivery, installation and 3-yearly maintenance by scaffolding platform with short distances for all transport. 20 year life for turbine (20 years output)
- **Option 2 long**: Delivery, installation and 3-yearly maintenance by scaffolding platform with long distances for all transport. 20 year life for turbine (20 years output)
- **Option 3 short**: Delivery and installation with short distances for all transport, no maintenance. 10 year life for turbine (10 years output)
- **Option 3 long**: Delivery and installation with long distances for all transport, no maintenance. 10 year life for turbine (10 years output)

Figure 2 shows how the impacts of these options (expressed as Ecopoints) compare over the service life (20 years for options 1 and 2, 10 years for option 3) for Micro-wind system 1.

To show the sensitivity of these impacts to the frequency of maintenance, Figure 3 shows the same information for annual rather than 3-yearly maintenance. Taking these figures and calculating the impact of the turbine, its installation and 3-yearly maintenance averaged over the period of output (i.e. 20 years for Options 1 and 2 and 10 years for Option 3) gives the results shown in Table 1.

These figures are used in Section 5 to estimate the annual electricity generation required to balance the embodied CO_2 (see Table 13).

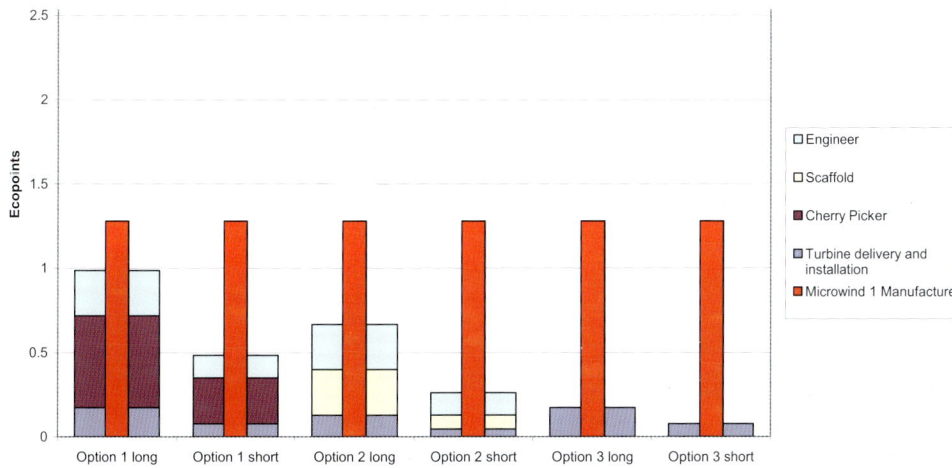

Figure 2. Delivery, installation and maintenance options for Micro-wind system 1 with 3-yearly maintenance for options 1 and 2.

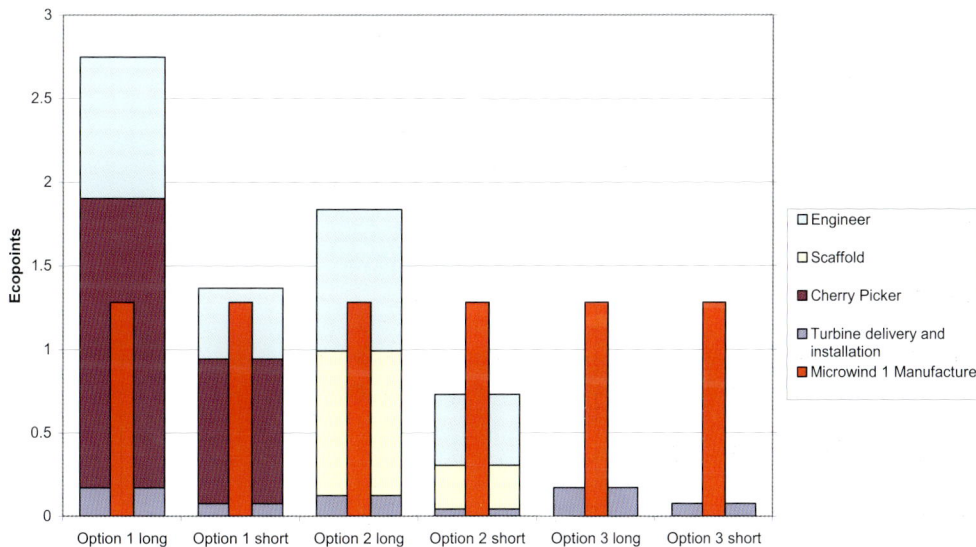

Figure 3. Delivery, installation and maintenance options for Micro-wind system 1 with annual maintenance for options 1 and 2.

Table 1: Annual environmental impact (averaged over expected life)					
Micro-wind system	System 1		System 2		System 3
	Ecopoints/year	kg CO_2/year	Ecopoints/year	kg CO_2/year	kg CO_2/year
Option 1 long	0.113	28.4	0.158	34.3	91.6
Option 1 short	0.088	18.5	0.133	24.4	81.7
Option 2 long	0.097	22.0	0.142	27.9	85.2
Option 2 short	0.077	14.0	0.122	20.0	77.2
Option 3 long	0.145	24.8	0.235	36.7	151.2
Option 3 short	0.136	21.0	0.226	32.9	147.4

3   ESTIMATION OF TYPICAL URBAN WIND RESOURCE

3.1   INTRODUCTION

3.1.1   Wind energy and power

Wind is air in motion, which flows at a certain velocity V. The air, although light, has mass m and as such has an energy content known as kinetic energy. The kinetic energy of the wind is referred to in this report as wind energy WE and can be expressed as:

$$WE = \frac{1}{2}\rho \times V^3 \times A \times t$$

where ρ = density of air (~ 1.225 kg/m³)

$\quad V$ = wind velocity

$\quad A$ = cross-sectional area of the wind stream

$\quad t$ = period of time over which the energy content is calculated

Power (P) is energy per unit of time and therefore the power content in the wind can be expressed as:

$$P = \frac{1}{2}\rho \times V^3 \times A$$

The equation for wind energy shows a cubic relation of the energy (power) content in the wind with the wind velocity. This has the following implications:

- A slight change in the wind speed drastically modifies the amount of power in the wind. For instance, a 25% increase in the wind speed would almost double the power available in the wind (see Figure 4) and conversely a 25% decrease in wind speed would more than halve the power available.
- Two sites with the same average wind speed can yield a different annual wind energy. This is due to the fact that the average of the cube of different wind speeds is always greater than the cube of the average wind speed.

Therefore, on evaluating the wind resource at a certain location it will be important not only to look at the average wind speed but also to undertake an analysis of the content of wind energy over a year.

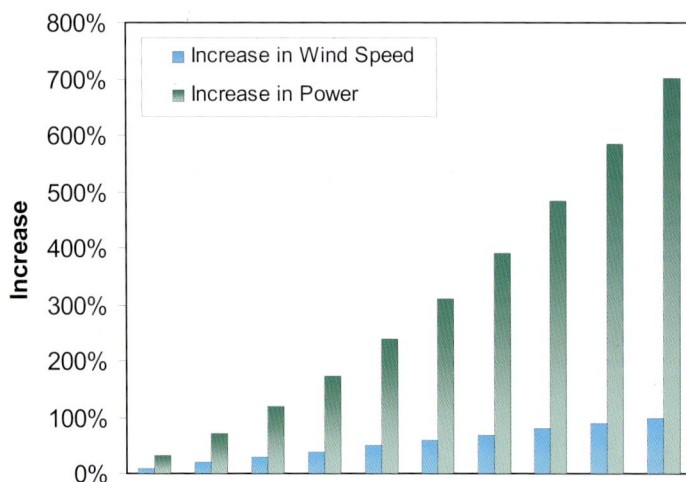

Figure 4. Relation of the power available in the wind with the wind speed.

3.1.2   Wind speed distribution

The amount of time that the wind blows within each speed range forms the wind speed distribution. This can also be shown as the number of hours that wind speeds, within a certain range, will blow over a year. Figure 5 shows the wind speed distribution over a year at the three representative locations (Manchester, Portsmouth and Wick), for which the wind resource is studied in this report. Data used have been provided by the Met Office for the Met Station locations shown in Appendix B. Ten-year measurements have been averaged to produce an average year. A frequency distribution for 30° wind direction intervals is given in 1 knot wind speed 'bins'. Figure 6 shows the Met Office data for Portsmouth, including the number of hours that the wind blows within each wind speed bin for each approaching wind direction interval. 0° refers to north, 90° to east, 180° to south and 270° to westerly wind directions.

   The importance of the wind speed distribution when undertaking a wind energy analysis can be understood if one considers that a few occurrences of high wind speeds will be translated in a high amount of energy over a year. This is due to the cubic relation between wind speed and wind energy.

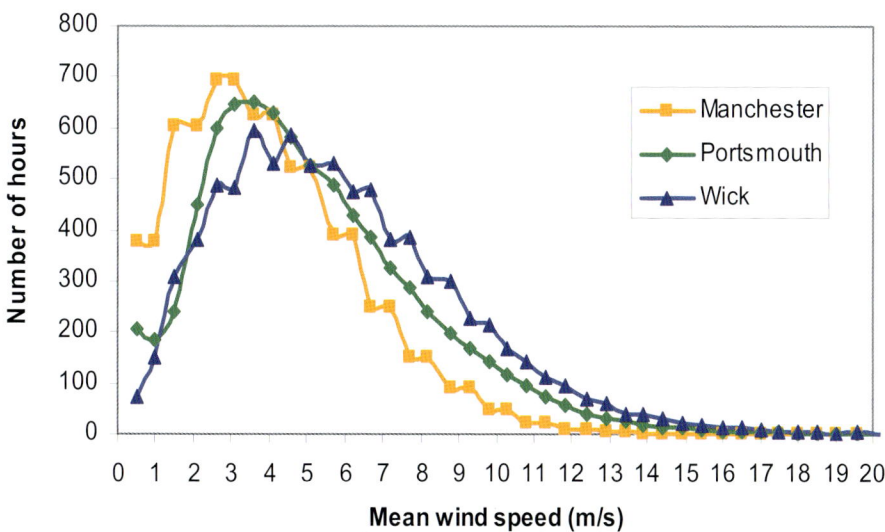

Figure 5. Annual wind speed distribution for three locations (source: Met Office).

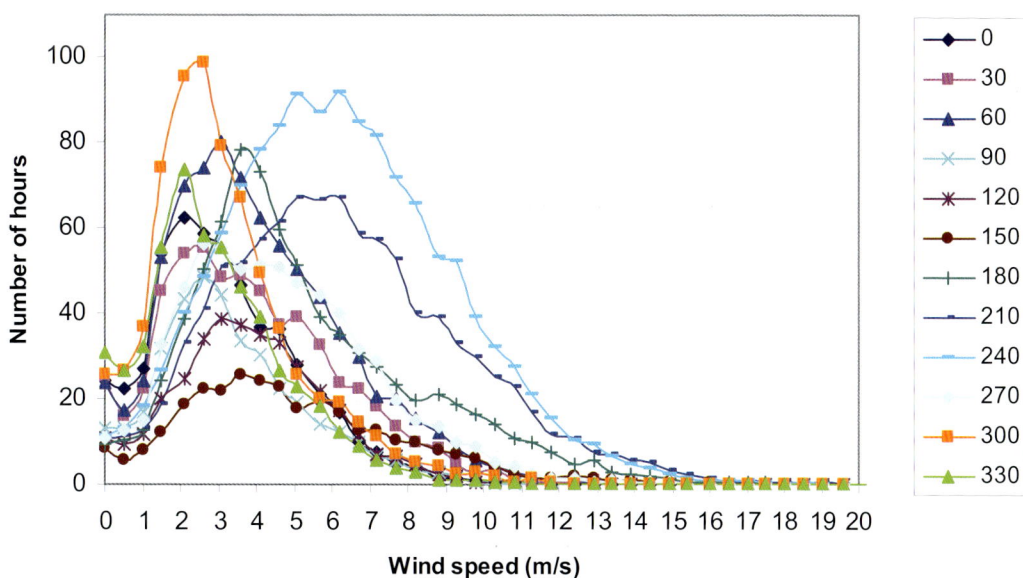

Figure 6. Annual wind speed distribution for Portsmouth for different wind directions (source: Met Office).

Figure 7 shows the wind energy over a year for different wind speeds. This figure gives an indication of the wind speeds which contain most of the energy over the year. It can be seen that the wind in Wick has the highest energy content, and that this is contained in relatively high wind speeds. Manchester however, will show most of its annual wind energy for lower wind speed regimes.

Different wind speed (energy) distributions suit different wind turbine types. The cut-in wind speed of a wind turbine is the wind speed above which the turbine starts producing useful power. The cut-out wind speed is the wind speed above which the turbine stops producing any useful power. Different manufacturers design turbines with cut-in and cut-out wind speeds. For low wind speed locations (such as Manchester), it will be important that the wind turbine starts producing power for low wind speeds, typically between 2.5 and 3.5 m/s. For locations with high occurrences of high wind speeds (such as Wick) it is important that the turbine is able to produce significant power for high wind speeds, as those contain a higher proportion of the wind energy content over a year.

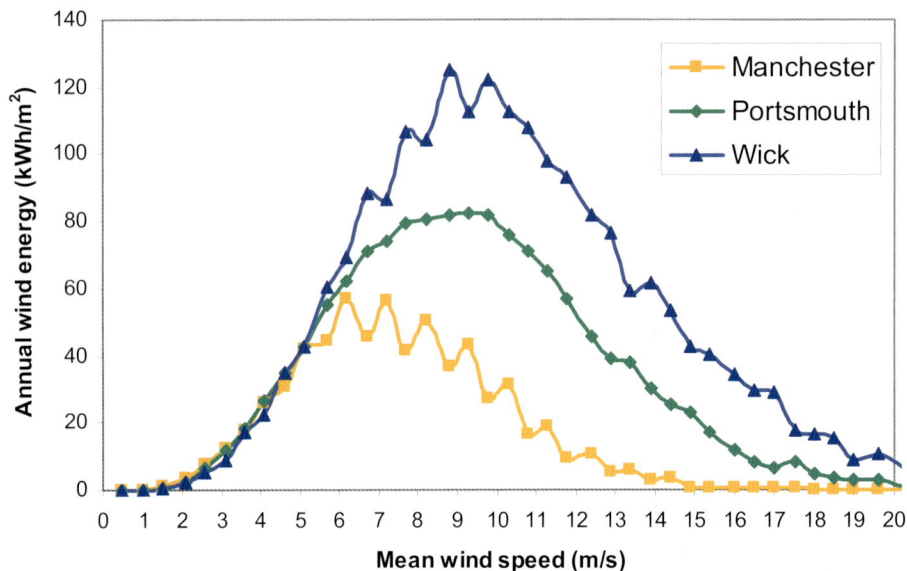

Figure 7. Annual wind energy versus wind speed for three locations (source: Met Office).

3.1.3   Wind roses

Wind turbines are traditionally installed on high towers and away from nearby obstacles. This implies that, in most situations, a wind turbine will be able to capture the wind from all wind directions.

However, for urban areas the wind resource available at particular locations will depend on the wind direction, as a wind turbine will normally be shaded by nearby obstacles. It is therefore important to evaluate the impact of nearby obstacles. This can be qualitatively assessed by the observation of *wind roses*. A wind rose, for the purpose of wind energy analysis, shows the energy content of the wind from different wind directions over a year. In order to maximise the turbine's output over a year, it is advisable that the wind turbine is relatively well exposed to the predominant winds.

Figures 8–10 show the wind rose for the three locations examined in this work.

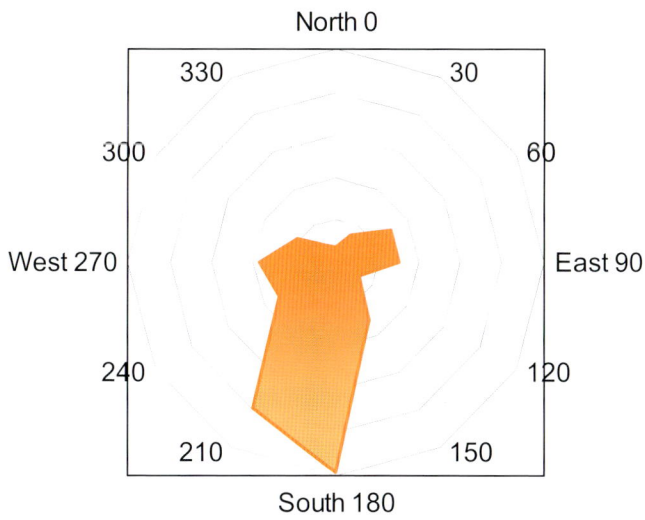

Figure 8. Energy content over a year for the different wind directions near Manchester.

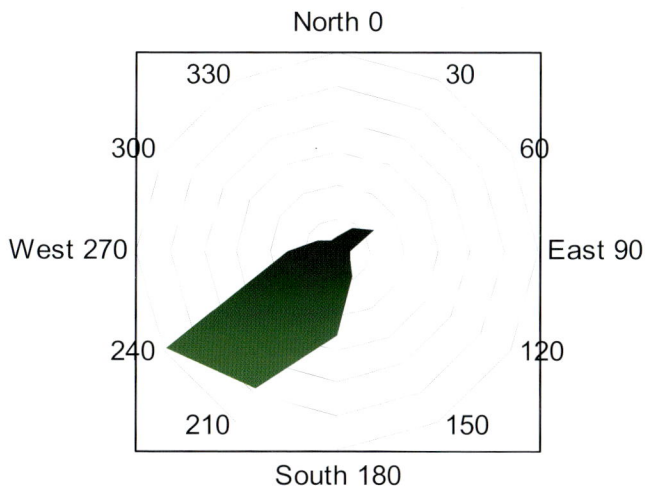

Figure 9. Energy content over a year for the different wind directions near Portsmouth.

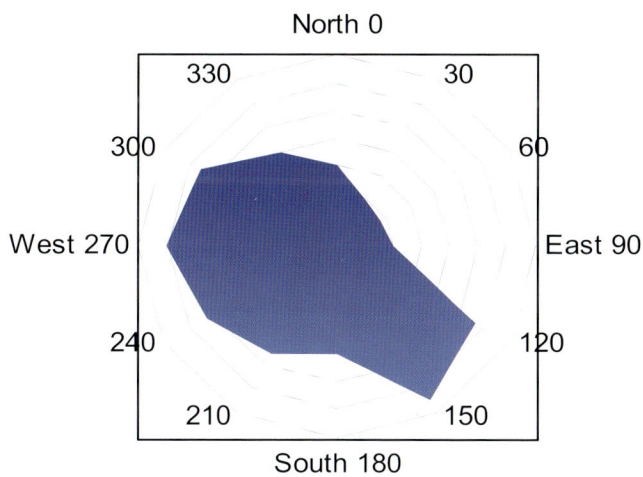

Figure 10. Energy content over a year for the different wind directions near Wick.

3.2   WIND RESOURCE – ADJUSTMENT FACTORS FOR URBAN ENVIRONMENTS
3.2.1   Wind scaling factors for urban environments
3.2.1.1   Methodology

In this study wind speed and power scaling factors were calculated that can be applied to meteorological data in order to allow for the aerodynamic roughness in urban areas.

The wind resource in urban areas was analysed using BREVe software [10] to establish the effect of local topographic roughness on wind speed.

BREVe is based on BS 6399-2 Code of practice for wind loads [11], and uses a UK-wide database of geographic locations, roughness and topography, together with knowledge of distance from the coast and assumptions about characteristic dimensions of buildings within each town. It calculates the aerodynamic roughness at any location, and the consequent effect on wind speeds.

BREVe produces data for exceptional storm events and the output used in this study is the mean hourly wind speed with a 1 in 50 year probability. In order to compare Met Office data with BREVe outputs, all wind speeds used in the calculations are given for 10 m height above ground level.

The urban wind velocity scaling factors were obtained using the following process:
- Data from three Met Office weather stations was obtained for Manchester, Portsmouth and Wick (Table 2)
- Grid references at five points within and around each city were selected using topographic maps (see Appendix B and Table 3)
- The BREVe model was run for the Met Office weather station locations and each of the five locations in the three cities
- Each velocity scaling factor (S_v) was calculated by taking the ratio of the BREVe wind speed at the Met Office station to the BREVe wind speed at each location in the city. This was done for each 30° of wind direction, giving 12 scaling factors per location, $S_v(\theta)$.

It is assumed that the 1 in 50 year event mean hourly speeds and mean annual wind speeds are attenuated to the same degree by the urban roughness elements.

For a specific location, the approaching wind will see different terrain roughness according to the surrounding landscape characteristics. BREVe considers roughness variations in 30° intervals and so velocity scaling factors for each wind direction have been calculated as:

$$S_v(\theta) = \frac{V_c(\theta)}{V_o(\theta)}$$

where $S_v(\theta)$ =   wind velocity scaling factor for wind blowing from direction θ

$V_c(\theta)$ =   1 in 50 year hourly mean speed attenuated by the roughness of the terrain, as determined by BREVe for wind blowing from direction θ

$V_o(\theta)$ =   1 in 50 year hourly mean speed at the Met Office station locations, as determined by BREVe for wind blowing from direction θ

As explained above, wind speeds are translated into power content by a cubic law. Therefore, it is possible to define a power scaling factor $S_P(\theta)$:

$$S_P(\theta) = \frac{V_c^3(\theta)}{V_o^3(\theta)}$$

These scaling factors have been obtained for the locations within each city shown in Table 3. Appendix B shows the city locations and the location of the associated Met station.

Table 2: Location of Met Office weather stations relative to cities.	
Manchester	South (Ringway Airport)
Portsmouth	East (Thorney Island)
Wick	North (Wick Airport)

Table 3: City locations for the estimation of the urban wind resource.
South east edge
North west edge
South east edge
North east edge
Centre of town

The scaling factors for velocity and power in the wind are shown in Figures 11–13 for Manchester, Portsmouth and Wick, respectively. The graphs show the factors for each approaching wind direction, where 0 refers to north, 90 to east, 180 to south and 270 to westerly winds. The wind from each direction is affected differently by the topographic roughness along its particular path. Differences within each city can also be appreciated, with the town centres always yielding the lowest wind resource, which is well below the wind speed values given by the Met Office weather stations.

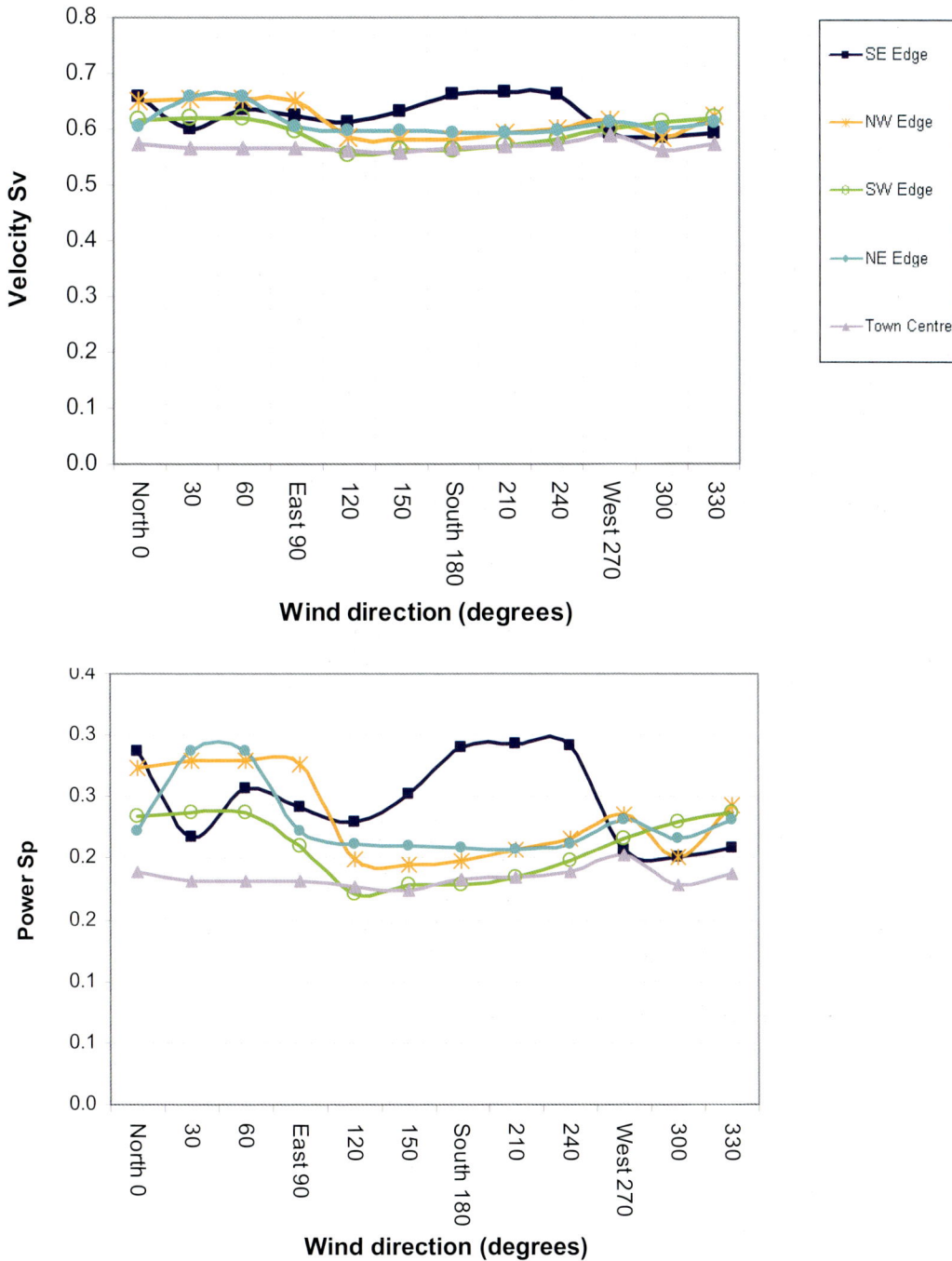

Figure 11. Scaling factors for velocity (S_V) and power (S_P): Manchester

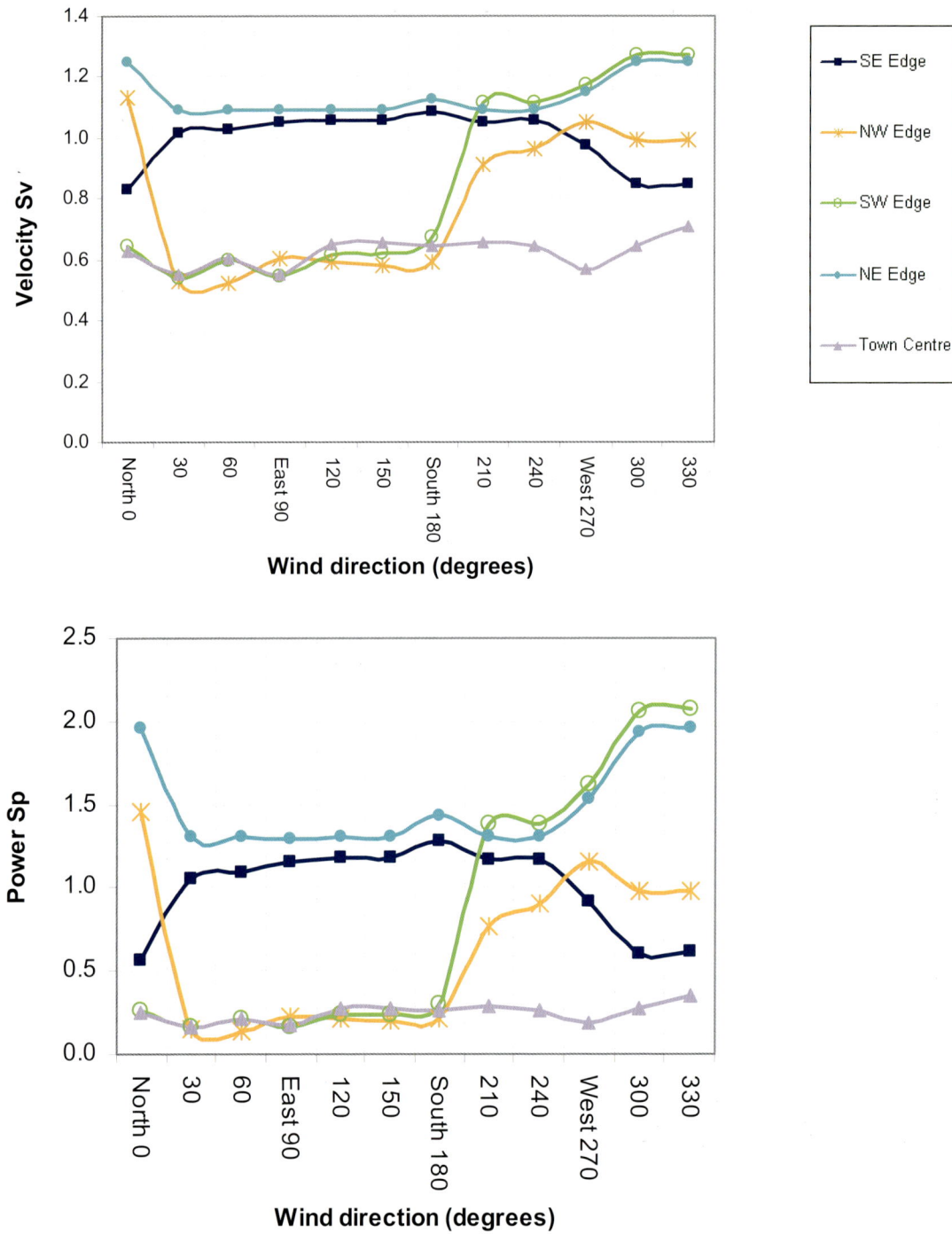

Figure 12. Scaling factors for velocity (S_V) and power (S_P): Portsmouth.

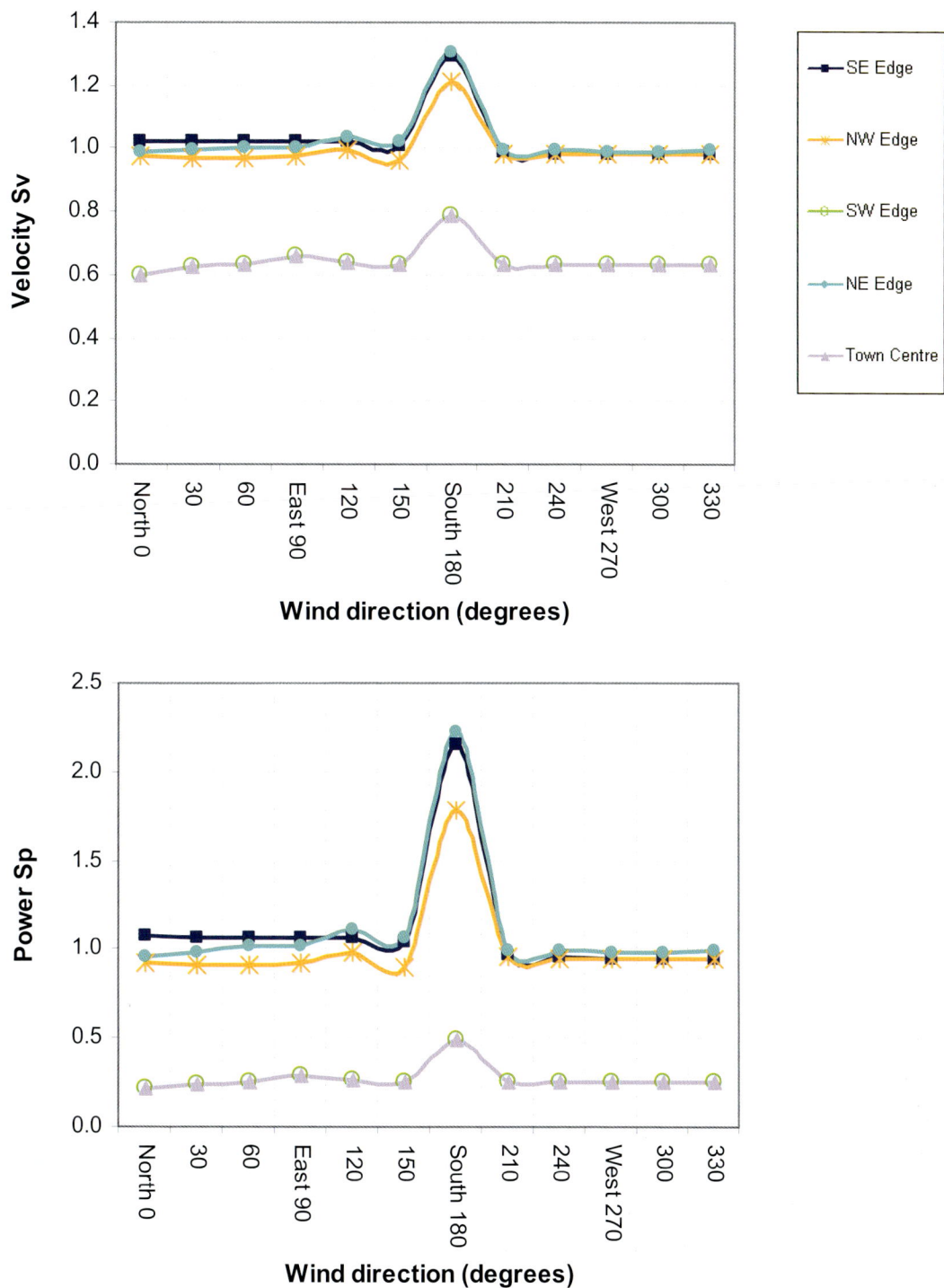

Figure13. Scaling factors for velocity (S_V) and power (S_P): Wick.

3.2.2   Influence of urban roughness in the annual wind resource

To show the effect of the urban terrain on the wind resource, the scaling factors for each location have been applied to the Met station data and the results are shown in Figures 14–16, and Tables 4–6. The figures depict the mean annual wind velocity and the available wind energy over a year at six locations in Manchester, Portsmouth and Wick. The tables and graphs were constructed using the following calculations:

The Met frequency data was used to derive the mean wind velocity over the observation period (usually 10 years) for each direction (θ) as follows:

$$\overline{V}_{MET}(\theta) = \sum_{i=1}^{N_{BIN}} \frac{n_i(\theta) \times V_i}{N_{OBS}(\theta)}$$

where
N_{BIN} = number of intervals of velocity used in the Met measurements ('bins')
$N_{OBS}(\theta)$ = total number of observations in wind direction θ at the Met station
$n_i(\theta)$ = frequency of occurrences of velocity V_i in direction θ at the Met station
V_i = mid point value of the velocity bin i measured at the Met station

$\overline{V}_{MET}(\theta)$ may be taken as representative of the mean annual velocity at the Met station for each direction.

The mean annual wind speed for each direction at any chosen location is therefore:

$$\overline{V}_{LOC}(\theta) = \overline{V}_{MET}(\theta) \times S_V(\theta)$$

Where $S_V(\theta)$ is the scaling factor for the particular direction θ, as described above.

The mean annual wind speed (all directions) for each chosen location in town V_{LOC} can be found by averaging over the wind directions:

$$\overline{V}_{LOC} = \sum_{\theta=0}^{\theta=330} \frac{N_{OBS}(\theta)}{N_{OBS}} \times \overline{V}_{LOC}(\theta)$$

where N_{OBS} is the total number of observations. The sum is executed over the 12 intervals of 30° given in the Met Office data.

The annual wind energy available at the locations WE_{LOC} was calculated in a similar way, but using the cube of the wind speeds and the power scaling factor S_P, along with number of hours of observation to convert to energy:

$$\overline{WE}_{LOC} = \frac{1}{2}\rho \sum_{\theta=0}^{\theta=330} S_P(\theta) \times \sum_{i=1}^{i=N_{bin}} h_i(\theta) \times V^3_i$$

where h_i = annual number of hours of velocity V_i in direction θ at the Met station
ρ = density of air (~1.225 kg/m³)

The reduction in the mean annual wind speed ∇_V and wind energy ∇_{WE} over a year due to terrain roughness is given in Tables 4–6 for Manchester, Portsmouth and Wick, respectively. This information is obtained directly from the reduction factors:

$$\nabla_V (\%) = (1 - S_V) \times 100$$

$$\nabla_{WE} (\%) = (1 - S_P) \times 100$$

Note: Negative values for ∇_V and ∇_{WE} indicate a wind resource higher than the one yielded at the associated Met station.

3.2.2.1   Large inland city: Manchester

Manchester can be considered to be representative of a large inland city. It shows predominant southerly winds over a year. For Manchester the predicted wind speed does not vary much within the town boundary (the exact location of the boundary is a subjective judgement in any case). This is due to the fact that all locations examined are surrounded by built-up areas (see Appendix B). A clear correlation between the aerodynamic roughness due to buildings and the wind resource in town can be seen from Figure 14 and Table 4.

A significant decrease in the wind speed can be seen for the different city locations compared to the velocity values given by the Met Office, with reductions around 40%. Due to the cubic relation of the wind power with the wind speed, this is translated to a decrease in the annual wind energy available of around 80%.

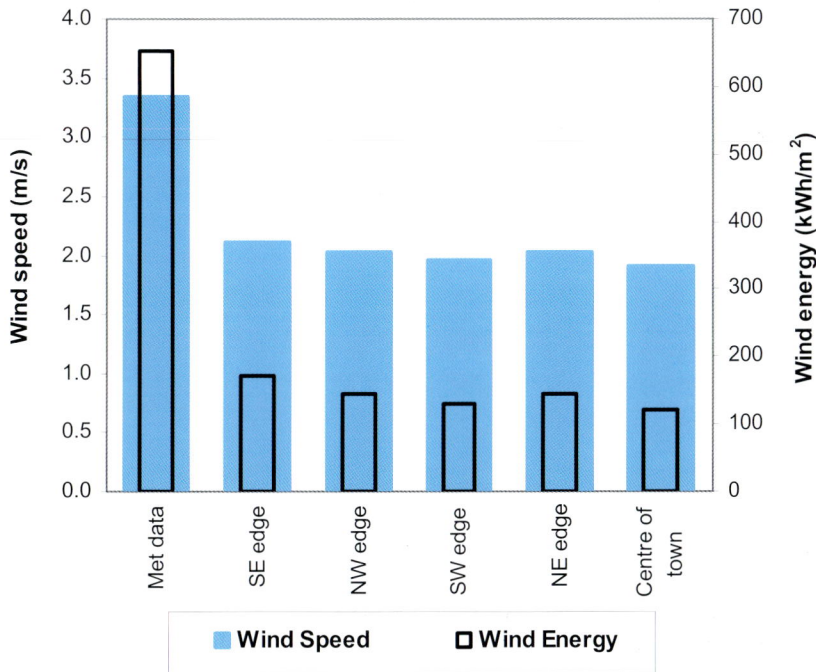

Figure 14. Influence of urban roughness on the annual wind resource. Manchester.

Table 4: Influence of urban roughness on the annual wind resource in Manchester.				
Location	Mean windspeed (m/s)	Mean windspeed reduction ∇_V (%)	Wind energy (kWh/m^2)	Wind energy reduction ∇_{WE} (%)
Met data	3.3		653	
South east edge	2.1	37%	172	74%
North west edge	2.0	40%	145	78%
South west edge	2.0	42%	129	80%
North east edge	2.0	40%	144	78%
Centre of town	1.9	43%	120	82%

3.2.2.2 Coastal city I: Portsmouth

Portsmouth is a coastal city in the south of England. It shows predominant south westerly winds over a year. For Portsmouth, the predicted wind speed varies significantly depending on the locations considered. Strangely, the northeast location yields the highest wind resource. This is surprising as Portsmouth shows predominantly south westerly wind directions and one would expect the location at the southwest edge of the city to yield a higher wind resource. The results, however, are consistent with the BREVe scaling factors shown in Figure 15. The reason for this is uncertain.

The city centre shows the lowest wind resource followed by the north west. The southern edge shows the same wind resource as that at the Met weather station. The north east, as mentioned, shows the highest wind resource of all the city locations studied with an mean annual wind speed 13% above the one yielded by the weather station.

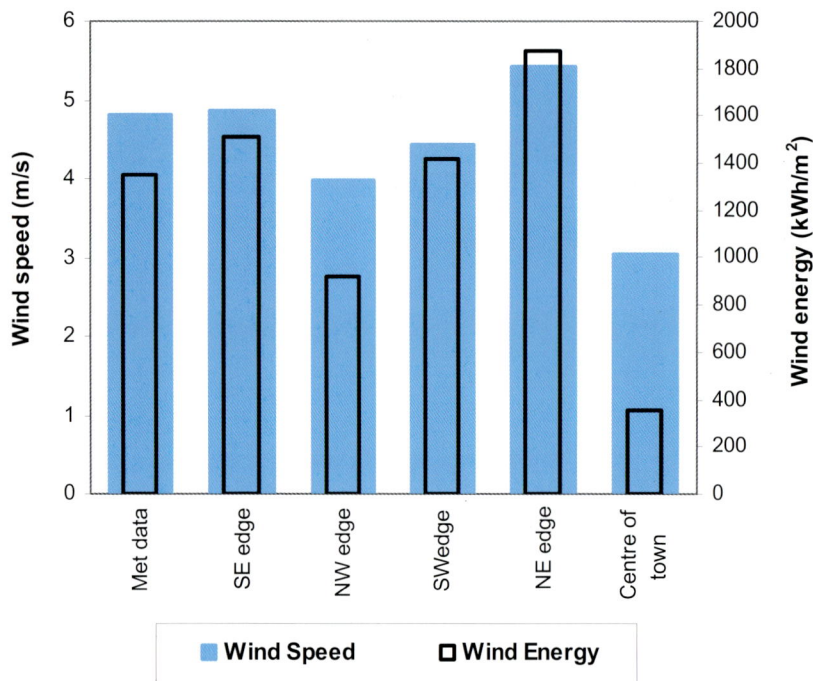

Figure 15. Influence of urban roughness on the annual wind resource. Portsmouth.

Table 5: Influence of urban roughness on the annual wind resource in Portsmouth.			
Location	Mean windspeed (m/s)	Mean windspeed reduction ∇_V (%)	Wind energy (kWh/m^2)	Wind energy reduction ∇_{WE} (%)
Met data	4.8		1354	
South east edge	4.9	-1%	1509	-11%
North west edge	4.0	17%	917	32%
South west edge	4.4	8%	1415	-4%
North east edge	5.4	-13%	1878	-39%
Centre of town	3.0	37%	350	74%

3.2.2.3 Coastal town II: Wick

Wick is a costal city in the north east of Scotland. It shows north westerly and south easterly predominant winds. The available wind resource at different locations in the city centre is closely related to the predominant wind directions as explained below.

The north west location yields the same amount of wind resource as the Met station. South east and north east locations yield the highest resource, slightly above the levels indicated by the Met station. This is understandable as both locations are exposed to the predominant wind directions.

City centre and southwest locations yield considerably lower resource than at the Met station, showing 35% lower wind speeds, which translates to around 70% less wind energy available over the year.

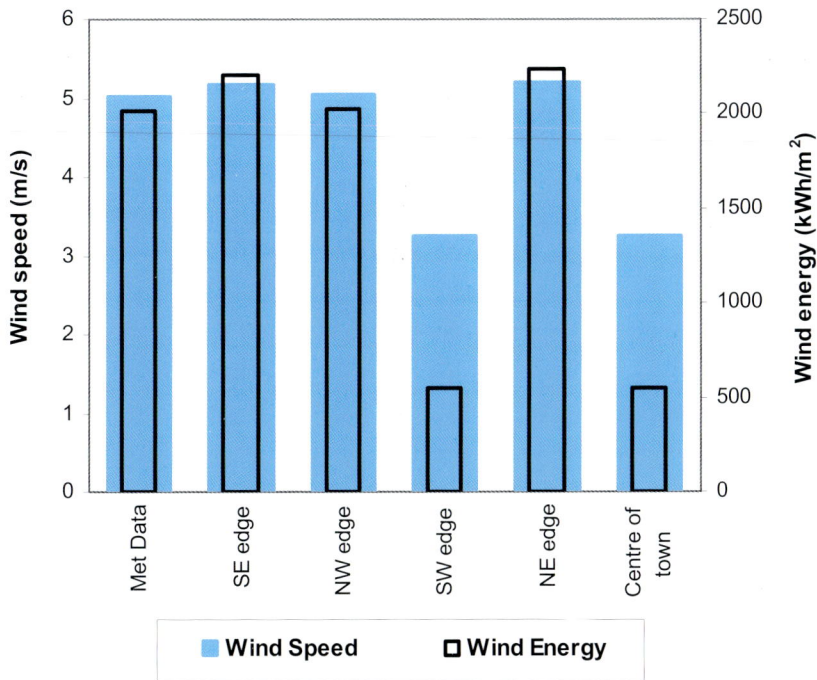

Figure 16. Influence of urban roughness on the annual wind resource. Wick..

Location	Mean windspeed (m/s)	Mean windspeed reduction ∇_V(%)	Wind energy (kWh/m^2)	Wind energy reduction ∇_{WE} (%)
Met data	5.0		2019	
South east edge	5.2	-3%	2204	-9%
North west edge	5.0	0%	2030	-1%
South west edge	3.2	35%	548	73%
North east edge	5.2	-4%	2236	-11%
Centre of town	3.2	35%	548	73%

Table 6: **Influence of urban roughness on the annual wind resource in Wick.**

sions – urban wind resource

UK locations examined, the city centres show wind speeds which are
40% less than those indicated by Met station readings, equivalent to around
energy over a year. Of the three cities examined, Manchester city centre
t wind resource, reflecting the large size of the city compared to Portsmouth
and Wick.

Other locations in the cities show a high variability in the annual wind resource,
mainly influenced by the roughness that the approaching wind sees for different wind
directions and the characteristics of the predominant winds. For those city locations most
exposed to the predominant winds, such as northeast and southeast Wick, the wind
resource has been observed to be at least as high as the resource at the Met station.

3.2.3   Use of the NOABL database for predicting the wind resource around built-up areas
3.2.3.1   NOABL – BERR wind speed database

The Department for Business, Enterprise and Regulatory Reform (BERR) (formerly
Department of Trade and Industry) has published a wind speed database known as the
NOABL database [12]. Wind speed values are given for 10, 25 and 40 m heights (the 25 m
NOABL map is shown in Appendix A).

The BERR description of NOABL is as follows:

*"The Department of Trade and Industry wind speed database contains estimates of the
annual mean wind speed throughout the UK. The data is the result of an air flow model
that estimates the effect of topography on wind speed. There is no allowance for the
effect of local thermally driven winds such as sea breezes or mountain/valley breezes. The
model was applied with 1 km resolution and makes no allowance for topography on a
small-scale or local surface roughness (such as tall crops, stone walls, or trees), both of
which may have a considerable effect on the wind speed. The data can only be used as a
guide and should be followed by on-site measurements for a proper assessment"* [12]

Therefore, the NOABL database does not take into account any feature other than the
general terrain topography.

3.2.3.2   Comparison between Met weather station data and the NOABL database

ETSU [13] described how the NOABL database was created. The database, also referred to as
the DTI wind speed database [12], got its name from the NOABL computer model developed
for quantifying wind flow over complex terrain. Actual data measured at UK Met stations in
the period 1975–84 were used in the model.

The NOABL wind speed database was created to allow a reasonable estimation of
wind speed for potential sites for large turbines and wind farms. It was created to remove
the influence that local topographic features may have on the wind resource, including tall
crops, trees and other obstacles. Therefore, the data given by the Met stations for locations
with terrain characteristics other than short grass was scaled up to compensate for local
roughness. To illustrate this, Figure 17 shows a comparison between the mean wind
velocity given by the Met Office and the NOABL database for the three cities examined in
this report.

As can be seen for all three locations, the Met Office data yield considerably lower
wind speeds than those indicated by the NOABL database. As outlined above, the main
reason for this is that the NOABL dataset considers the whole UK roughness to be a cut
grass terrain, and omits the effect of trees, forest, tall crops, etc.

Furthermore, Figure 18 shows a comparison between the city centre wind speeds
scaled down by the BREVe factors and the NOABL results. The differences are even more
significant.

It is concluded that, using NOABL wind speeds for estimating the wind resource in a
built-up area is highly misleading. Although it is possible that local effects due to other
buildings can sometimes enhance the wind speed, in most situations and for built-up
areas, the wind speed will be severely reduced, which will result in considerably
overestimating the resource available.

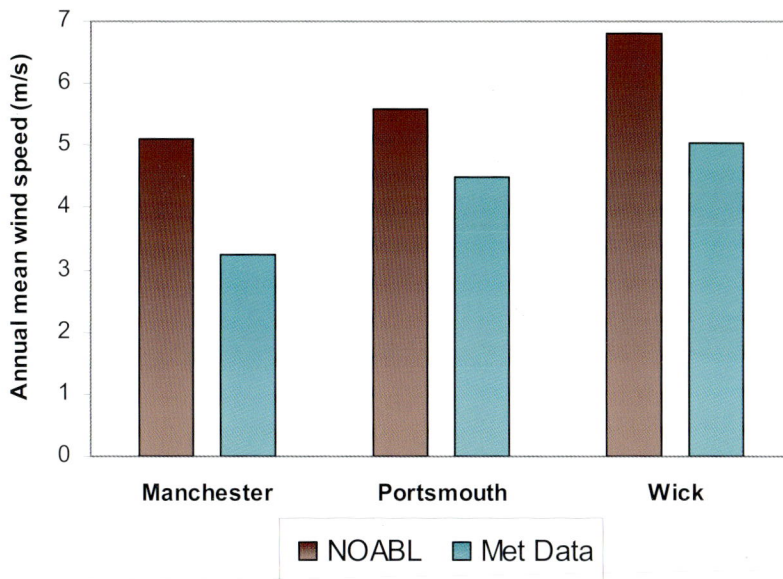

Figure 17. Comparison of the annual mean wind speed from the NOABL database and Met weather station data.

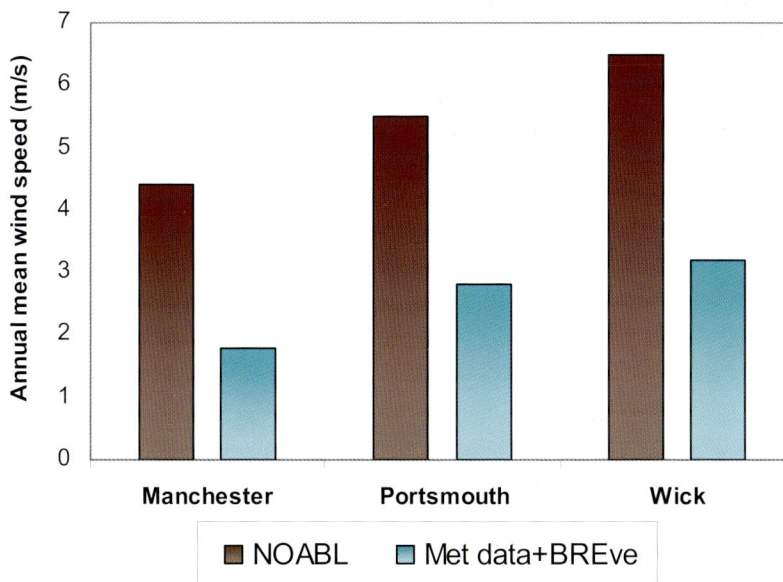

Figure 18. Comparison of the annual mean wind speed from the NOABL database and the calculated wind resource for the urban locations.

3.2.3.3 Conclusions

Manufacturers of turbines normally present information on annual electricity production for different wind speeds. Simply using this information with the NOABL database in locations affected by local obstacles and topographic features will lead to large overestimates of the wind resource.

It is suggested that a modified NOABL database that accounts for the UK terrain roughness is developed. This could be done by combining the information provided by the BREVe software and its database of UK terrain roughness with the NOABL database to create a corrected wind speed database that accounts for the local topography of the UK terrain. This database would be more suitable for estimating the wind resource for small turbine installations for different terrain types, from open farmland to suburban and urban areas. This would be a useful tool for more realistic estimation of the wind resource for small wind energy installations. It could serve as a guideline for installers under the new UK Microgeneration Certification scheme and wind energy engineers in general. However, this work is outside of the scope of this report.

4   ELECTRICITY GENERATION BY BUILDING-MOUNTED WIND TURBINES IN TYPICAL URBAN SCENARIOS

4.1   INTRODUCTION

As part of this project, wind tunnel testing was undertaken for a series of building types in order to estimate the effect that some typical domestic buildings have on the approaching wind. The results of this work are published separately [3]. In this work, mean velocity ratios were obtained, which relate the undisturbed approaching wind speed at a height of 10 m to the velocity at different positions above the building roof. These ratios can be directly applied to the velocity values given in Section 3.2.2 to account for the influence that the building has in modifying the approaching wind speed and allow the wind resource seen by building mounted wind turbines to be estimated.

In this section, estimates are made of the annual electricity generation for the three micro-wind turbine systems with a number of different building mounting positions for several typical urban locations. The equivalent dimensions of the buildings examined by Blackmore [3] are 6 m height to the eaves with plan dimensions of 10 × 10 m and 25° and 50° roof pitches, which are representative of typical detached houses. Blackmore [3] showed that the results for a terraced house of the same geometry are very similar and therefore the results presented here may be used for terraced and detached houses.

4.2   METHODOLOGY FOR THE ELECTRICITY CALCULATION

There are several methodologies for calculating (estimating) the electrical energy produced by a wind turbine. Here, the power curve method has been used. This method uses the wind turbine power curve along with the wind speed distribution of a site (see Figure 19).

The power curve of a wind turbine relates the electrical power output of the turbine to wind speed. The wind speed distribution relates the frequency of occurrence, or number of hours, of different wind speeds over a given period. If the number of hours of each wind speed is known, and the power curve of a turbine is provided, it is possible to estimate the electricity produced by the turbine over a year. This is illustrated in Figure 19.

This study shows simple electricity calculations using different manufacturers' power curves and Met Office weather data scaled to allow for:

- the effect of the urban roughness, as described in Section 3.2
- the effect of the building on the local wind speed, using the wind tunnel mean velocity ratios given by Blackmore [3].

Although the conditions under which the turbine power curves were obtained is, in most cases, uncertain, it is expected that they were obtained in wind tunnels or using free-standing mast-mounted turbines, far away from obstacles and in areas of low turbulence.

The wind turbines considered in this work were all horizontal axis, axial flow turbines (HAWTs), which are characterised by a horizontal axis of rotation. In order to maximise the amount of wind intercepted by the rotor, the turbine uses a mechanism to align the rotor perpendicularly to the wind. Another type of turbine, vertical axis wind turbines (VAWTs), have the axis of rotation perpendicular to the ground. Unlike HAWTs, VAWTs do not need to be oriented to the wind direction, and may therefore be better suited to the urban environment. However, they are less efficient than HAWTs and their benefits in urban areas are yet to be experimentally proven.

HAWTs are more widely available than HAWTs and represent a more established technology. They are also cheaper at present and are therefore more applicable in studies such as this on micro-wind turbines in urban areas.

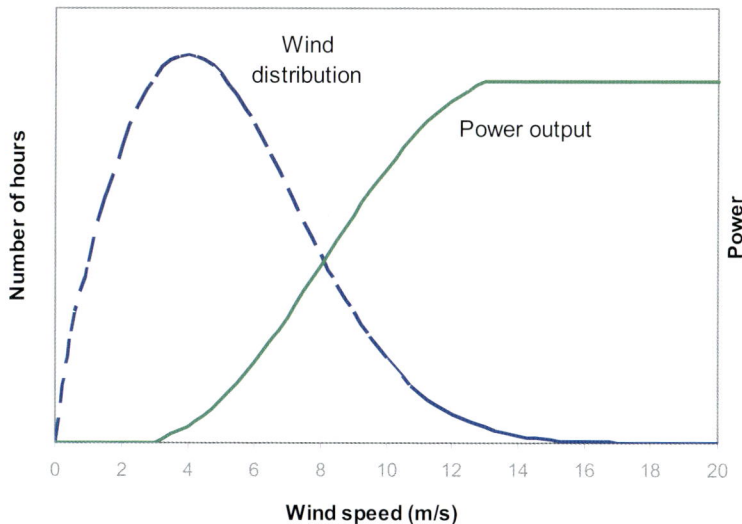

Figure 19. Wind speed distribution and generic wind turbine power curve, illustrating the power curve method for calculating the electricity produced by a wind turbine over a year.

Micro and small HAWTs usually employ an end tail system in order to align the wind turbine with the wind direction. When the wind direction is inconsistent, the turbine will be continually realigning on its axis. This is known as *yawing*. The high turbulence levels around buildings can affect the yaw behaviour of the turbine and can reduce its power output. This is the case if a turbine is not able to follow sudden changes in the wind direction.

This effect should be the object of further research and therefore the results presented here should be interpreted with care. It is possible that the electricity generated will be lower once the turbulence in the wind is taken into account. This effect is not considered in this report. Monitoring of full-scale installations will give a better insight into this effect.

4.3   RESULTS

Estimates of the electricity generated annually by the three micro-wind turbine systems have been made for three locations, Manchester, Portsmouth and Wick.

These were chosen as they provide a range of wind conditions due to their varied location (see Appendix A), their local geography and the extent of the built-up urban area. Manchester is a large inland city with relatively low wind speeds and can be considered representative of many large cities in low lying areas of the Midlands and South east. Portsmouth is a medium sized city with moderate wind speeds and Wick is a small town in a windy location.

The building types considered are shown in Table 7. Surrounding buildings were taken into account during the wind tunnel tests [3] by using models of 10 m high buildings at a fixed density around the building model studied.

Because of the many possible permutations of the mounting position and height, the estimates of the annual electricity generation have been made for the for the three micro-wind turbine systems at mounting positions 1, 7, 13 and 15 as detailed by Blackmore [3] and as shown in Figure 20 for the building oriented to the west

Table 7: Types of building and surrounding layout.	
Label	**Description**
Det DP 25	Detached house, 25° dual-pitch roof, isolated building
Det DP 25 (10 m)	Detached house, 25° dual-pitch roof, surrounding buildings at 10 m
Det DP 25 (20 m)	Detached house, 25° dual-pitch roof, surrounding buildings at 20 m
Det DP 25 (50 m)	Detached house, 25° dual-pitch roof, surrounding buildings at 50 m
Det DP 50	Detached house, 50° dual-pitch roof, isolated building
Det DP 50 (10 m)	Detached house, 50° dual-pitch roof, surrounding buildings at 10 m
Det DP 50 (20 m)	Detached house, 50° dual-pitch roof, surrounding buildings at 20 m
Det DP 50 (50 m)	Detached house, 50° dual-pitch roof, surrounding buildings at 30 m

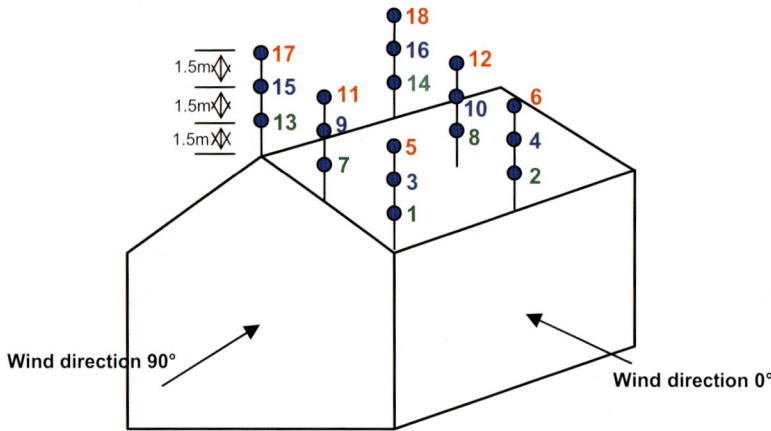

Figure 20. Turbine mounting positions chosen for the wind tunnel experiments [1].

The mounting positions 1, 7 and 13, which are 1.5 m above three points on the gable end of the building, are considered to be typical and relatively practical mounting positions giving adequate clearance for the turbine blades. Mounting position 15 (3 m above the ridge at the gable end of the building) is included to investigate the sensitivity to positioning in terms of height but is considered to be challenging in terms of structural integrity for typical domestic buildings. Mounting positions 4.5 m above roof height are considered unlikely for domestic buildings.

To allow comparison, the first two rows of data in Tables 8–12 (shaded grey) show, respectively, the electricity that might be expected from the turbine mounted on a 10 m mast:

- in clear air at the Met station (i.e. outside the city) and
- in clear air at the selected location with the wind speeds scaled for the urban terrain but without local building effects. (It should be noted for some of the building mounted positions the actual height is greater than 10 m)

4.3.1   Manchester

Table 8 shows the results for Manchester, on the southwest edge of the city (see Appendix B) for the various building types given in Table 7 in an urban terrain. It can be seen from Section 3.2.2.1 that the wind resource was very similar for all of the city locations studied and hence these results can be considered as representative of most areas in the city.

Table 8. Annual electricity output (kWh) for micro-wind turbines in Manchester (south west edge).

Micro turbine	3 m above ridge (pos. 15)			1.5 m above ridge (pos. 13)			1.5 m above mid roof (pos. 7)			1.5 m above eaves (pos. 1)		
	1	2	3	1	2	3	1	2	3	1	2	3
Met data	614	367	698	614	367	698	614	367	698	614	367	698
Scaled for terrain	91	21	80	91	21	80	91	21	80	91	21	80
Det DP 25	123	37	114	120	35	108	94	27	87	78	21	71
Det DP 25 (10 m)	54	9	46	24	2	20	22	2	18	9	0	8
Det DP 25 (20 m)	57	13	50	24	3	20	20	2	16	14	1	12
Det DP 25 (50 m)	43	7	37	32	4	27	28	3	23	26	4	22
Det DP 50	145	46	133	147	47	135	70	19	64	64	19	59
Det DP 50 (10 m)	105	27	92	63	13	54	26	3	22	12	1	11
Det DP 50 (20 m)	98	27	89	84	24	77	23	3	19	12	1	10
Det DP 50 (50 m)	106	33	97	67	15	60	27	3	23	17	2	15

4.3.2   Portsmouth

It can be seen from Section 3.2.2.2 that the wind resource is much greater on the outskirts of the city than in the city centre. Results are therefore given for two locations in Portsmouth for the various building types given in Table 7 in an urban terrain. Table 9 shows the results for the north east edge of the city, which has the highest wind resource, and Table 10 shows the results for the city centre, which has the lowest wind resource.

Table 9: Annual electricity output (kWh) for micro-wind turbines in Portsmouth (north east edge).

Micro turbine	3 m above ridge (pos. 15)			1.5 m above ridge (pos. 13)			1.5 m above mid roof (pos. 7)			1.5 m above eaves (pos. 1)		
	1	2	3	1	2	3	1	2	3	1	2	3
Met data	1136	838	1489	1136	838	1489	1136	838	1489	1136	838	1489
Scaled for terrain	1390	1131	2001	1390	1131	2001	1390	1131	2001	1390	1131	2001
Det DP 25	1589	1367	2471	1595	1404	2532	1487	1268	2257	1325	1103	1982
Det DP 25 (10 m)	996	728	1303	537	330	633	436	283	536	253	152	292
Det DP 25 (20 m)	810	610	1130	502	323	609	429	277	514	351	214	406
Det DP 25 (50 m)	730	518	953	695	462	854	609	393	735	544	370	676
Det DP 50	1731	1563	2869	1714	1571	2810	1258	1080	1922	1218	1039	1848
Det DP 50 (10 m)	1486	1237	2203	1063	800	1436	600	394	731	317	197	375
Det DP 50 (20 m)	1321	1061	1920	1085	861	1591	552	357	665	341	195	381
Det DP 50 (50 m)	1156	954	1768	1080	829	1491	616	409	753	471	289	550

Table 10: Annual electricity output (kWh) for micro-wind turbines in Portsmouth (city centre).

Micro turbine	3 m above ridge (pos. 15)			1.5 m above ridge (pos. 13)			1.5 m above mid roof (pos. 7)			1.5 m above eaves (pos. 1)		
	1	2	3	1	2	3	1	2	3	1	2	3
Met data	1136	838	1489	1136	838	1489	1136	838	1489	1136	838	1489
Scaled for terrain	298	139	307	298	139	307	298	139	307	298	139	307
Det DP 25	416	230	462	434	243	481	376	201	408	320	167	347
Det DP 25 (10 m)	186	73	182	85	27	79	77	29	74	39	10	35
Det DP 25 (20 m)	182	87	189	84	30	80	67	21	63	52	14	47
Det DP 25 (50 m)	141	58	141	112	38	106	93	30	88	85	29	80
Det DP 50	509	302	581	487	285	553	320	172	346	300	158	322
Det DP 50 (10 m)	351	179	374	201	85	202	98	32	91	48	15	44
Det DP 50 (20 m)	306	155	327	259	135	284	83	24	75	46	10	40
Det DP 50 (50 m)	303	166	338	216	95	219	98	31	91	63	15	55

4.3.3 Wick

It can be seen from Section 3.2.2.3 that the wind resource is much greater on the outskirts of the town than in the town centre. Results are therefore given for two locations in Wick for the various building types given in Table 7 in an urban terrain. Table 11 shows the results for the south east edge of the city, which has the highest wind resource, and Table 12 shows the results for the city centre, which has the lowest wind resource.

Table 11: Annual electricity output (kWh) for micro-wind turbines in Wick (south east edge).

Micro turbine	3 m above ridge (pos. 15)			1.5 m above ridge (pos. 13)			1.5 m above mid roof (pos. 7)			1.5 m above eaves at (pos. 1)		
	1	2	3	1	2	3	1	2	3	1	2	3
Met data	1542	1230	2173	1542	1230	2173	1542	1230	2173	1542	1230	2173
Scaled for terrain	1639	1359	2396	1639	1359	2396	1639	1359	2396	1639	1359	2396
Det DP 25	1779	1552	2760	1771	1527	2715	1507	1241	2208	1159	919	1663
Det DP 25 (10 m)	1218	917	1629	664	417	784	527	322	617	295	162	324
Det DP 25 (20 m)	1064	812	1454	640	411	769	560	369	682	442	279	523
Det DP 25 (50 m)	957	693	1249	867	604	1108	758	525	966	660	482	882
Det DP 50	1937	1751	3138	1923	1716	3085	972	780	1428	852	689	1271
Det DP 50 (10 m)	1724	1468	2596	1278	991	1766	718	493	899	395	231	449
Det DP 50 (20 m)	1572	1330	2358	1345	1119	1989	666	476	872	439	277	523
Det DP 50 (50 m)	1488	1259	2268	1303	1035	1848	724	521	951	566	397	743

Micro turbine	3 m above ridge (pos. 15)			1.5 m above ridge (pos. 13)			1.5 m above mid roof (pos. 7)			1.5 m above eaves (pos. 1)		
	1	2	3	1	2	3	1	2	3	1	2	3
Met data	1542	1230	2173	1542	1230	2173	1542	1230	2173	1542	1230	2173
Scaled for terrain	489	259	529	489	259	529	489	259	529	489	259	529
Det DP 25	604	349	679	592	359	662	466	249	504	343	178	369
Det DP 25 (10 m)	320	144	322	138	42	127	107	34	100	55	12	48
Det DP 25 (20 m)	284	133	295	135	46	127	120	42	112	89	26	81
Det DP 25 (50 m)	234	98	232	209	81	202	177	69	174	169	75	171
Det DP 50	716	438	829	700	426	805	307	167	338	279	155	310
Det DP 50 (10 m)	550	307	606	352	170	365	166	64	161	78	20	69
Det DP 50 (20 m)	506	276	551	417	223	454	162	67	161	93	30	87
Det DP 50 (50 m)	490	277	546	375	188	398	180	76	180	143	62	145

Table 12: Annual electricity output (kWh) for micro-wind turbines in Wick (city centre).

4.4   CONCLUSIONS

For the typical systems modelled there is a very wide range of predicted annual electricity generation, from 0 to 3138 kWh/year. Not surprisingly, the lowest output is for a turbine mounted just above the eaves in a densely built-up area in a large city (Manchester) and the highest output is for a larger turbine mounted 3 m above the ridge of an isolated building on the outskirts of a small town in a windy location (Wick).

The maximum output for Manchester is only 147 kWh/year and it can be seen that the outputs are considerably lower where the effects of surrounding buildings have been taken into account. In these cases, the output only exceeds 100 kWh/year if the turbine is mounted 3 m above the ridge, which is not considered practical for most domestic buildings. These results are for a location in the south west edge of the city but the data shown in Section 3.2.2.1 suggests that the wind resource is very similar throughout the city and hence similar results can be expected anywhere in Manchester.

The data in Sections 3.2.2.2 and 3.2.2.3 show that, for Portsmouth and Wick, there are significant differences between the wind resource on the outskirts of the cities from that in the centres. Results have therefore been given for outskirt and city centre locations.

For Portsmouth the output ranges from 152 to 2869 kWh/year on the north east edge and from 10 to 581 kWh/year in the city centre. The figures for Wick are higher: 162 to 3138 kWh/year on the north east edge and 12 to 829 kWh/year in the town centre. In all these cases there is a significant reduction in output when considering the more practical situations where there are surrounding buildings within 50 m and the turbine is mounted only 1.5 m above the roof.

It should be noted that these results are considered to be optimistic as the calculations do not take account of turbulence effects. Also, although the general effect of the town terrain and surrounding buildings is accounted for, it is assumed that the building on which the turbine is mounted is not overshadowed by larger buildings, trees or other obstructions. It is well recognised qualitatively that high levels of turbulence reduce the output (see Section 4.2) but there are no quantitative data that can be used to account for the effect in the calculations. Blackmore [3] shows that the turbulence intensity is greater for the lower mounting positions and that it is increased by the presence of surrounding buildings. The effects of turbulence are therefore likely to make the reductions observed for the more practical mounting positions mentioned above even greater.

5 CO_2 PAYBACK FOR DOMESTIC MICRO-WIND TURBINES IN URBAN ENVIRONMENTS

The electrical outputs for the micro-wind turbines shown in Section 4 are calculated in kWh per year. BRE hold data for the production of electricity within the UK that is sourced from the ecoinvent database produced by the Swiss Government [6]. This includes all the impacts associated with the extraction and transport of fuels, production of electricity and distribution to the consumer. This gives an impact of 0.632 kg CO_2 equivalent per kWh of electricity.

For assessing the CO_2 payback, the output from the micro-wind turbines is assumed to replace electricity sourced by the consumer from the low voltage grid (220–240 V) and the impact of this has been used to model the 'benefit' of micro-wind generation.

It should be noted that the ecoinvent data provides higher greenhouse gas emissions than some other models of electricity generation which could be used for this purpose. The reasons for this are that:

1   It includes all greenhouse gas emissions, not just CO_2.
2   All upstream impacts are modelled, including transport
3   Low voltage electricity is modelled, including the losses associated with the consumer network, rather than 'average electricity' which includes large amounts of high and medium voltage electricity, which is consumed by industry with reduced losses.
4   Data is based on production in 2000 rather than more recent data.
5   Data is calculated using the same methodology as the data for manufacture, transport and maintenance and so is directly comparable.
6   Data is not the 'marginal' electricity, i.e. the actual type of electricity generation that it is assumed would be reduced due to the use of micro-wind generation. Marginal electricity can be calculated over the long or short term.

For these reasons, it is likely that the BRE calculation provides a best case, as the impact of conventional grid generation is maximised and therefore has maximum benefit when compared to the impact of manufacture, installation and maintenance. However, some studies do recommend the use of marginal electricity models based on the replacement of coal-fired generation with CO_2 impact figures of about 0.800 kg/kWh.

The impact of electricity used within the study is: 1 kWh = 0.00276 Ecopoints and 0.632 kg CO_2 equivalent.

Taking the information provided in Section 2, and these figures for the impact of electricity generation, the amount of output per year needed to balance the impact of manufacture, installation and maintenance can be calculated for the Micro-wind systems (see Table 13).

Table 13. Annual electrical generation required to balance embodied CO_2 for the different installation and maintenance options.

	Micro-wind system 1		Micro-wind system 2		Micro-wind system 3	
	kg CO_2 per year	kWh/year required to balance CO_2	kg CO_2 per year	kWh/year required to balance CO_2	kg CO_2 per year	kWh/year required to balance CO_2
Option 1 long	28.4	45	34.3	54	91.6	145
Option 1 short	18.5	29	24.4	39	81.7	129
Option 2 long	22.0	35	27.9	44	85.2	135
Option 2 short	14.0	22	20.0	32	77.2	122
Option 3 long	24.8	39	36.7	58	151.2	239
Option 3 short	21.0	33	32.9	52	147.4	233

From this it can be seen that, for Micro-wind systems 1 and 2, Option 1 long requires the greatest annual output to balance the embodied CO_2 and Option 2 short requires the least. However, for Micro-wind system 3, Option 3 (with no maintenance) requires considerably more annual output to balance the embodied CO_2 than the other options. This is because for Option 3 the service life is assumed to be only 10 years and the cost of System 3 relative to the maintenance costs is much greater than for Systems 1 and 2.

The CO_2 payback periods have been calculated for Option 2 short and Option 1 long. For practical purposes these are considered to be, respectively, the best and worst cases for the maintenance options for all of the systems, as having invested in the larger turbine (System 3) it would be sensible to conduct three-yearly maintenance to avoid reducing the service life.

The CO_2 payback is achieved when the impact saved by the electricity generated (expressed as kg CO_2 equivalent) is equal to the impact of manufacturing, installation and maintenance (also expressed as kg CO_2 equivalent). The time taken to payback T_P can be calculated using the equation:

$$T_P = (M + I) / (G_a \times E - A)$$

Where
T_P	=	payback time in years
M	=	environmental impact of manufacture, expressed as kg of CO_2 equivalent
I	=	environmental impact of installation, expressed as kg of CO_2 equivalent
G_a	=	annual electricity generated in kWh per year
E	=	impact of electricity generation, expressed as kg of CO_2 per kWh ($= 0.632$)
A	=	annual share of the environmental impact of the 3-yearly maintenance, expressed as kg of CO_2 per year

Tables 14–18 show the CO_2 payback times calculated for the three micro-wind turbine systems in all of the locations and conditions used in Section 4 for the calculation of the electricity generation. For each set of conditions, the payback periods (in years) have been calculated for the best and worst case maintenance options, as explained above, and are shown as a range between these two values. The payback periods are shown in years (rounded to one decimal place or two significant places), and 'n/p' is shown where there is no payback (or the payback period is more than 100 years). For example, '4.1–14' indicates that the payback period is between 4.1 and 14 years depending on the maintenance option used, '17–n/p' indicates that the period is between 17 years and no payback, and 'n/p' indicates that there is no payback for either condition.

In these tables the cells are highlighted as follows:
- RED indicates that there is no CO_2 payback within the expected life of the system (20 years);
- ORANGE indicates that there may be a CO_2 payback within the expected life depending on the maintenance option chosen, but the payback period exceeds the expected life for the worst case option;
- YELLOW indicates that the CO_2 payback period may exceed 10 years (i.e. exceeds 10 years for the worst case maintenance option).

Again, to allow comparison, the first two rows of data in Tables 14–18 (shaded grey) show, respectively, the payback periods that might be expected from the turbine mounted on a 10 m mast:
- in clear air at the Met station (i.e. outside the city) and
- in clear air at the selected location with the wind speeds scaled for the urban terrain but without local building effects. *(It should be noted that, for some of the building-mounted positions, the actual height above ground is greater than 10 m)*

Table 14: CO_2 payback periods (years) for micro-wind turbines in Manchester (south west edge).

Micro turbine	3 m above ridge (pos. 15)			1.5 m above ridge (pos. 13)			1.5 m above mid pitch (pos. 7)			1.5 m above eaves (pos. 1)		
	1	2	3	1	2	3	1	2	3	1	2	3
Met data	0.5–0.7	1.4–1.7	3.4–3.6	0.5–0.7	1.4–1.7	3.4–3.6	0.5–0.7	1.4–1.7	3.4–3.6	0.5–0.7	1.4–1.7	3.4–3.6
Scaled for terrain	3.7–6.2	36–n/p	32–46	3.7–6.2	36–n/p	32–46	3.7–6.2	36–n/p	32–46	3.7–6.2	36–n/p	32–46
Det DP 25	2.7–4.1	17–65	22–28	2.8–4.3	18–87	23–30	3.6–5.9	26–n/p	29–41	4.4–7.8	26–n/p	37–56
Det DP 25 (10 m)	6.7–15	n/p	60–n/p	19–n/p	n/p	n/p	22–n/p	n/p	n/p	n/p	n/p	n/p
Det DP 25 (20 m)	6.3–14	87–n/p	54–n/p	19–n/p	n/p	n/p	25–n/p	n/p	n/p	48–n/p	n/p	n/p
Det DP 25 (50 m)	8.8–26	n/p	78–n/p	13–97	n/p	n/p	15–n/p	n/p	n/p	17–n/p	n/p	n/p
Det DP 50	2.9–4.1	13–32	18–23	2.2–3.3	13–30	18–22	5.0–9.4	43–n/p	41–67	5.5–11	43–n/p	45–78
Det DP 50 (10 m)	4.1–6.3	25–n/p	27–37	5.6–11	92–n/p	49–92	16–n/p	n/p	n/p	66–n/p	n/p	n/p
Det DP 50 (20 m)	4.5–6.9	25–n/p	28–39	4.1–7.0	27–n/p	33–48	20–n/p	n/p	n/p	66–n/p	n/p	n/p
Det DP 50 (50 m)	4.1–6.2	19–n/p	26–35	5.2–10	65–n/p	44–76	16–n/p	n/p	n/p	31–n/p	n/p	n/p

Table 15: CO_2 payback periods (years) for micro-wind turbines in Portsmouth (north east edge).

Micro turbine	3 m above ridge (pos. 15)			1.5 m above ridge (pos. 13)			1.5 m above mid pitch (pos. 7)			1.5 m above eaves (pos. 1)		
	1	2	3	1	2	3	1	2	3	1	2	3
Met data	0.3–0.4	0.6–0.7	1.6–1.6	0.3–0.4	0.6–0.7	1.6–1.6	0.3–0.4	0.6–0.7	1.6–1.6	0.3–0.4	0.6–0.7	1.6–1.6
Scaled for terrain	0.2–0.3	0.5–0.5	1.2–1.2	0.2–0.3	0.5–0.5	1.2–1.2	0.2–0.3	0.5–0.5	1.2–1.2	0.2–0.3	0.5–0.5	1.2–1.2
Det DP 25	0.2–0.3	0.4–0.4	0.9–1.0	0.2–0.3	0.4–0.4	0.9–1.0	0.2–0.3	0.4–0.5	1.0–1.1	0.2–0.3	0.5–0.5	1.2–1.2
Det DP 25 (10m)	0.3–0.4	0.7–0.8	1.8–1.9	0.6–0.8	1.6–1.9	3.7–4.0	0.7–1.0	1.8–2.3	4.4–4.7	1.3–1.7	3.5–4.7	8.1–9.1
Det DP 25 (20m)	0.4–0.5	0.8–1.0	2.1–2.2	0.6–0.8	1.6–2.0	3.8–4.1	0.7–1.0	1.8–2.3	4.6–4.9	0.9–1.2	2.4–3.1	5.8–6.3
Det DP 25 (50m)	0.4–0.6	1.0–1.2	2.4–2.6	0.5–0.6	1.1–1.3	2.7–2.9	0.5–0.7	1.3–1.6	3.2–3.4	0.6–0.8	1.4–1.7	3.5–3.7
Det DP 50	0.2–0.2	0.3–0.4	0.8–0.8	0.2–0.2	0.3–0.4	0.8–0.9	0.3–0.3	0.5–0.6	1.2–1.3	0.3–0.3	0.5–0.6	1.3–1.3
Det DP 50 (10m)	0.2–0.3	0.4–0.5	1.1–1.1	0.3–0.4	0.6–0.8	1.6–1.7	0.5–0.7	1.3–1.6	3.2–3.4	1.0–1.4	2.6–3.4	6.3–6.9
Det DP 50 (20m)	0.2–0.3	0.5–0.6	1.2–1.3	0.3–0.4	0.6–0.7	1.5–1.5	0.6–0.8	1.4–1.8	3.5–3.8	0.9–1.3	2.7–3.5	6.2–6.8
Det DP 50 (50m)	0.3–0.4	0.5–0.6	1.3–1.4	0.3–0.4	0.6–0.7	1.6–1.6	0.5–0.7	1.3–1.5	3.1–3.3	0.7–0.9	1.8–2.2	4.3–4.6

Table 16: CO_2 payback periods (years) for micro-wind turbines in Portsmouth (city centre).

Micro turbine	3 m above ridge (pos. 15)			1.5 m above ridge (pos. 13)			1.5 m above mid pitch (pos. 7)			1.5 m above eaves (pos. 1)		
	1	2	3	1	2	3	1	2	3	1	2	3
Met data	0.3–0.4	0.6–0.7	1.6–1.6	0.3–0.4	0.6–0.7	1.6–1.6	0.3–0.4	0.6–0.7	1.6–1.6	0.3–0.4	0.6–0.7	1.6–1.6
Scaled for terrain	1.1–1.5	3.8–5.2	7.7–8.6	1.1–1.5	3.8–5.2	7.7–8.6	1.1–1.5	3.8–5.2	7.7–8.6	1.1–1.5	3.8–5.2	7.7–8.6
Det DP 25	0.8–1.0	2.3–2.9	5.1–5.5	0.7–1.0	2.1–2.7	4.9–5.3	0.9–1.1	2.6–3.4	5.8–6.3	1.0–1.3	3.1–4.2	6.8–7.5
Det DP 25 (10 m)	1.8–2.5	7.6–13	13–16	4.0–6.9	26–n/p	32–47	4.5–8.0	23–n/p	35–52	9.9–37	n/p	85–n/p
Det DP 25 (20 m)	1.8–2.6	6.3–9.9	13–15	4.1–7.0	22–n/p	32–46	5.2–10	36–n/p	42–69	7.0–16	70–n/p	58–n/p
Det DP 25 (50 m)	2.3–3.5	9.9–19	17–21	3.0–4.7	16–59	24–31	3.6–6.0	22–n/p	29–40	4.0–6.7	23–n/p	32–46
Det DP 50	0.6–0.8	1.7–2.1	4.0–4.3	0.7–0.9	1.8–2.3	4.2–4.6	1.0–1.4	3.0–4.0	6.8–7.5	1.1–1.4	3.3–4.5	7.4–8.1
Det DP 50 (10 m)	0.9–1.2	2.9–3.9	6.3–6.9	1.6–2.3	6.4–10	12–14	3.4–5.6	20–n/p	28–38	7.6–19	68–n/p	63–n/p
Det DP 50 (20 m)	1.1–1.4	3.4–4.6	7.2–8.0	1.2–1.7	3.9–5.4	8.4–9.4	4.1–7.1	30–n/p	34–51	8.2–22	n/p	71–n/p
Det DP 50 (50 m)	1.1–1.4	3.2–4.2	7.0–7.7	1.5–2.1	5.7–8.7	11–13	3.5–5.6	21–n/p	28–38	5.6–11	62–n/p	48–87

Table 17: CO$_2$ payback periods (years) for micro-wind turbines in Wick (north east edge).

Micro turbine	3 m above ridge (pos. 15)			1.5 m above ridge (pos. 13)			1.5 m above mid pitch (pos. 7)			1.5 m above eaves (pos. 1)		
	1	2	3	1	2	3	1	2	3	1	2	3
Met data	0.2–0.3	0.4–0.5	1.1–1.1	0.2–0.3	0.4–0.5	1.1–1.1	0.2–0.3	0.4–0.5	1.1–1.1	0.2–0.3	0.4–0.5	1.1–1.1
Scaled for terrain	0.2–0.2	0.4–0.4	1.0–1.0	0.2–0.2	0.4–0.4	1.0–1.0	0.2–0.2	0.4–0.4	1.0–1.0	0.2–0.2	0.4–0.4	1.0–1.0
Det DP 25	0.2–0.2	0.3–0.4	0.8–0.9	0.2–0.2	0.3–0.4	0.9–0.9	0.2–0.3	0.4–0.5	1.1–1.1	0.3–0.4	0.6–0.7	1.4–1.5
Det DP 25 (10 m)	0.3–0.3	0.6–0.7	1.4–1.5	0.5–0.6	1.2–1.5	3.0–3.2	0.6–0.8	1.6–2.0	3.8–4.1	1.1–1.5	3.2–4.3	7.3–8.1
Det DP 25 (20 m)	0.3–0.4	0.6–0.7	1.6–1.7	0.5–0.6	1.2–1.5	3.0–3.2	0.6–0.7	1.4–1.7	3.4–3.7	0.7–1.0	1.8–2.3	4.5–4.8
Det DP 25 (50 m)	0.3–0.4	0.7–0.9	1.9–2.0	0.4–0.5	0.8–1.0	2.1–2.2	0.4–0.5	1.0–1.2	2.4–2.6	0.5–0.6	1.1–1.3	2.6–2.8
Det DP 50	0.2–0.2	0.3–0.3	0.7–0.8	0.2–0.2	0.3–0.3	0.8–0.8	0.3–0.4	0.7–0.8	1.6–1.7	0.4–0.5	0.7–0.9	1.8–1.9
Det DP 50 (10 m)	0.2–0.2	0.3–0.4	0.9–0.9	0.3–0.3	0.5–0.6	1.3–1.4	0.4–0.6	1.0–1.3	2.6–2.8	0.8–1.1	2.2–2.9	5.2–5.7
Det DP 50 (20 m)	0.2–0.3	0.4–0.5	1.0–1.0	0.2–0.3	0.5–0.5	1.2–1.2	0.5–0.6	1.1–1.3	2.7–2.8	0.7–1.0	1.9–2.3	4.5–4.8
Det DP 50 (50 m)	0.2–0.3	0.4–0.5	1.0–1.1	0.2–0.3	0.5–0.6	1.3–1.3	0.4–0.6	1.0–1.2	2.5–2.6	0.6–0.7	1.3–1.6	3.1–3.4

Table 18: CO$_2$ payback periods (years) for micro-wind turbines in Wick (town centre).

Micro turbine	3 m above ridge (pos. 15)			1.5 m above (pos. 13)			1.5 m above mid pitch (pos. 7)			1.5 m above eaves (pos. 1)		
	1	2	3	1	2	3	1	2	3	1	2	3
Met data	0.2–0.3	0.4–0.5	1.1–1.1	0.2–0.3	0.4–0.5	1.1–1.1	0.2–0.3	0.4–0.5	1.1–1.1	0.2–0.3	0.4–0.5	1.1–1.1
Scaled for terrain	0.7–0.9	2.0–2.5	4.4–4.8	0.7–0.9	2.0–2.5	4.4–4.8	0.7–0.9	2.0–2.5	4.4–4.8	0.7–0.9	2.0–2.5	4.4–4.8
Det DP 25	0.5–0.7	1.5–1.8	3.4–3.7	0.5–0.7	1.5–1.9	3.5–3.8	0.7–0.9	2.1–2.6	4.7–5.0	0.9–1.3	2.9–3.9	6.4–7.0
Det DP 25 (10 m)	1.0–1.4	3.7–5.0	7.4–8.1	2.4–3.6	14–41	19–24	3.1–5.0	19–n/p	25–33	6.5–15	n/p	57–n/p
Det DP 25 (20 m)	1.1–1.5	4.0–5.5	8.1–9.0	2.5–3.7	13–32	19–24	2.8–4.3	14–41	22–28	3.8–6.4	26–n/p	31–45
Det DP 25 (50 m)	1.4–1.9	5.5–8.3	10–12	1.6–2.2	6.8–11	12–14	1.8–2.6	8.1–14	14–16	1.9–2.8	7.4–12	14–17
Det DP 50	0.4–0.6	1.2–1.4	2.8–3.0	0.5–0.6	1.2–1.5	3.0–3.1	1.0–1.4	3.1–4.2	7.0–7.7	1.2–1.6	3.4–4.6	7.6–8.5
Det DP 50 (10 m)	0.6–0.8	1.7–2.1	3.9–4.1	0.9–1.2	3.1–4.1	6.5–7.1	2.0–2.8	8.8–16	15–18	4.4–7.9	39–n/p	37–58
Det DP 50 (20 m)	0.6–0.8	1.9–2.3	4.3–4.6	0.8–1.0	2.3–3.0	5.2–5.6	2.0–2.9	8.3–15	15–18	3.6–6.0	22–n/p	29–41
Det DP 50 (50 m)	0.7–0.9	1.9–2.3	4.3–4.6	0.9–1.1	2.8–3.6	5.9–6.5	1.8–2.6	7.3–12	13–16	2.3–3.4	9.2–17	17–20

6   LIFE CYCLE COSTS AND FINANCIAL PAYBACK FOR MICRO-WIND TURBINES

6.1   INTRODUCTION TO LIFE CYCLE COSTING

OGC Procurement Guide 07 [14] defines the principles of whole life costs as follows:

> The whole life cost of a facility (often referred to as through life costs) are the costs of acquiring it (including consultancy, design, and construction costs, and equipment), the costs of operating it and the costs of maintaining it over its whole life through to its disposal – that is, the total ownership costs. These costs include: internal resources and departmental overheads, where relevant; they also include risk allowance as required; flexibility (predicted alterations for known change in business requirements, for example), refurbishment costs, and costs relating to sustainability and health and safety aspects.

Life cycle costing is defined in the International Standard ISO 15686 [15] as follows:

> The economic assessment considering all agreed projected significant and relevant cost flows over a period of analysis expressed in monetary value. The projected costs are those needed to achieve defined levels of performance, including reliability, safety and availability.

Although this definition captures a number of the essential issues in life cycle costing it is not particularly easy as an introduction, and it may be easier to consider the Whole Life Cost Forum definition [16], which is:

> The analysis of all relevant and identifiable financial cashflows regarding the acquisition and use of an asset.

None of these definitions capture the point that life cycle costing (LCC) is a tool, to assist in making choices between different options with different cash flows over a period of time. That is, LCC is a forward planning tool to anticipate such questions as:
- What needs doing to the building or component?
- When should it be done?
- How much will it cost?

Why use life cycle costing?

The benefits of LCC include:
- optimising the total cost of ownership/occupation by balancing initial capital and running costs
- providing data on actual performance and operation compared with predicted performance for use in planning and benchmarking
- promoting realistic budgeting for operation, maintenance and repair
- encouraging discussion and recording of decisions about the durability of materials and components at the outset of the project
- encouraging analysis of business needs and communication of these to the project team
- ensuring that risk and cost analysis of loss of functional performance due to failure or inadequate maintenance occur.

What activities should be considered when undertaking LCC at whole building level?

At whole building level, the life cycle cost is arrived at by adding up the initial capital cost of a building, and the inspection, operating and maintenance costs of all the individual components – windows, heating plant etc – over their useful lives, plus the replacement cost at the end of life. These costs are usually presented at today's costs by discounting the total sum.

What activities should be considered when undertaking LCC at component level?

At individual component level, the life cycle cost is the initial installation cost, either as new or as a replacement, plus the costs of inspecting, operating and maintaining it and its replacement cost when it wears out. Access costs such as scaffolding need to be included for most of these activities.

It is important to capture all the activities associated with the ownership of a component if it is to last its normal life expectancy. Maintenance is costly and is often put off because of budget constraints. One consequence of putting it off is that accelerated wear or deterioration can occur.

Facilities costs may be cyclical (for inspection, maintenance and related management aspects), or intermittent (for reactive maintenance). Replacement cycles for a component need to be predicted in order to estimate its replacement cost. Some costs may be treated as income such as capital allowances and sale.

When should repairs and maintenance be carried out?

The design life of the building will influence the number of times the elements that will not last for this period will need replacing. In housing, life cycle costs are usually considered over 60 years. During this time a number of components will need to be replaced several times as they reach the end of their working lives. Some elements are expected to last the life of the building including those that are difficult to replace – foundations, under building drainage, the main structure etc – whereas mechanical and electrical services will require replacing a number of times (Figure 21).

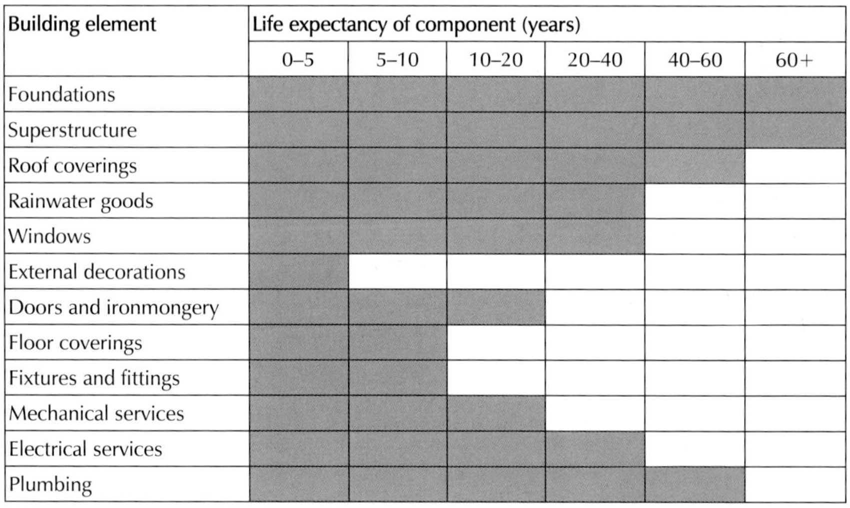

Building element	Life expectancy of component (years)					
	0–5	5–10	10–20	20–40	40–60	60+
Foundations	▓	▓	▓	▓	▓	▓
Superstructure	▓	▓	▓	▓	▓	▓
Roof coverings	▓	▓	▓	▓	▓	
Rainwater goods	▓	▓	▓	▓		
Windows	▓	▓	▓	▓	▓	
External decorations	▓					
Doors and ironmongery	▓	▓	▓			
Floor coverings	▓	▓				
Fixtures and fittings	▓	▓				
Mechanical services	▓	▓	▓			
Electrical services	▓	▓	▓	▓		
Plumbing	▓	▓	▓	▓	▓	

Figure 21. Examples of life expectancy.

A number of issues will affect life expectancy of a component – the amount of maintenance, excessive wear and tear, environmental conditions, its quality, and how well it was installed. A number of sources, including manufacturers, provide basic information on life expectancies but some levels of judgement are needed to arrive at a reasonably accurate prediction of life expectancy based on in-use situations.

6.2   WHAT COSTS ARE TAKEN INTO ACCOUNT WHEN UNDERTAKING LCC FOR A WIND TURBINE?

Capital cost

The capital cost is needed for two purposes. It is included in the overall total life cycle cost, and it is also used as an estimate for the costs of component replacements at the end of their working lives. The costs of initial construction will vary depending on whether the life cycle cost is at whole building level or just at component level, e.g. a wind turbine. At component level, the cost of installation and the purchase price are required, plus professional fees and the costs of statutory approvals if appropriate.

Maintenance

The following types of maintenance need to be considered at whole building and component levels: preventative, scheduled, corrective, condition based, reactive (emergency), predictive, and deferred. The following activities will be included:

- inspection
- monitoring
- testing
- condition surveys
- planning
- repairs
- replacements
- refurbishments.

Accuracy of costs

Values for the costs should be as accurate as possible. Greater effort may be required for the most significant cost variables. Values have been derived from a direct estimation from known costs and components.

How to present life cycle costing

The format for reporting life cycle costing is often the net present value (NPV) – which is a single figure representing all the future costs and incomes at their equivalent present value. For this exercise, we have used the discount rate of 3.5% as currently recommended under Treasury guidelines; where undiscounted costs are shown they are noted as being at zero discount.

Life cycle costing for micro-wind turbines

Information on costs and service lives was gathered by approaching manufacturers by telephone to inquire about cost, installation, servicing, maintenance contracts, accessibility, etc. Some information was difficult to obtain as the manufacturers do not necessarily sell the turbines directly to consumers and the companies selling the turbines do not always install them and did not have full control of the final cost to the consumer, so internet searches were carried out to supplement the information received.

The figures in Table 19 give a 60-year life cycle cost incorporating the price information received from each manufacturer. The cost of access to the turbines for installation and/or maintenance operations has been based on using a 7.5 m cherry picker for all systems.

A service life of 20 years has been allowed for all of the systems when 3-yearly maintenance has been included but a service life of 10 years has been allowed for the example with no servicing (see Section 2.7). All prices exclude VAT.

Table 19: Summary of LCC (60 year study period) using manufacturers'/installers' data.

	60-year costs (zero discount)		60-year net present value (3.5% discount)	
	Total cost	Annual cost	Total cost	Annual cost
Micro-wind system 1	£14,172	£236	£7,761	£129
Micro-wind system 2	£12,866	£214	£6,711	£112
Micro-wind system 3	£19,822	£330	£11,170	£186
Micro-wind system 2 (no maintenance)	£14,638	£244	£7,443	£124

Information from manufacturers
Micro-wind system 1
The capital cost of the unit is £2850 plus site survey and standard installation costs. No specific requirement for routine replacement of parts was suggested by the manufacturer other than noting that an annual service would be sensible. A service life of 15 to 20 years was suggested by the manufacturer.

Micro-wind system 2
The turbine cost is £1798 plus site survey and standard installation costs. No maintenance agreements or recommendations are currently in place. A 10-year life is suggested as a minimum based on the life suggested by the suppliers of the individual parts. The manufacturers oversee the assembly of the unit and have an installation company to install the turbines. They recommend an annual inspection to check for unusual events such as bird damage, etc. This would require access via a tower, scaffold or other means to elevate the inspector to the level of the turbine (we have assumed a cherry picker).

Micro-wind system 3
The cost of system 3 is £3500 plus site survey and standard installation costs. The manufacturers advised that a service life of 20 years is expected and although they have no definite servicing or maintenance requirements, we have assumed a service interval as other turbines. They were unable to provide either costs (because this work was done by installers), or installers details, so the price has been taken from 'Better Generation' web site [17].

The effects of maintenance
In order to show the effect of maintenance on the different turbines, in all the tables in this section we have included a life cycle cost for a turbine which assumes no maintenance taking place. This option is for a typical turbine based on Micro-wind system 2 as that is the system with the lowest overall capital cost. With no maintenance or inspection being carried out, we consider it prudent to limit the life to 10 years.

Inspection and access
As turbines will be fixed to various points on a domestic property, access methods may vary. However, as all turbines are assumed to be fixed to masts at high points on the building, access for inspection by ladder would be unsafe so scaffold tower or cherry picker type access would be necessary. In all cases we have allowed for cherry picker access as scaffold towers would be unsupported for the upper portion and cherry pickers would be able to access more different installation positions than towers. Little specific information was given by the manufacturers about maintenance, with requirements varying between 'no regular inspection' to believing that an annual inspection would be a 'good idea'. In order to compare turbines on a consistent basis and to include for reasonable servicing intervals, we have allowed for access by cherry picker for inspection at three year intervals. A cherry picker in these circumstances is only marginally more expensive than a scaffold tower (owing to the time taken to erect and dismantle a tower), but would allow quicker access so more inspections could be undertaken within a time period.

Electricity generation
In Section 4, details have been given for the generation of electricity per turbine at different locations across the country. This data has been used to calculate the value of the electricity to be set against the cost of the turbine and factored into the life cycle cost forecast. An average of costs per unit charged by electricity suppliers is 9.6365 p/kWh and this is the rate applied to the turbine generation.

Although the annual generation figures shown fall below the average electricity consumption for a 2–3 bedroom house (i.e. 4000–5500 kWh/year) it is likely that in practice some of the electricity generated will be sold back to the electricity supplier because the periods of high generation will not always match the periods of high demand. Most suppliers buy back at a lower rate than they supply (e.g. 4.5 p/unit) with some, under certain circumstances, buying back at the same rate as the selling rate. For the purposes of this LCC, it is assumed that the suppliers will buy back at the same cost as they sell electricity. This is the 'best case' and is used here to determine if there is a financial payback in the best case for the typical installations modelled.

Details of the likely electricity generation at the different locations considered are given in Section 4.3, but for the purposes of this LCC, we have based electricity generation value on the best situation in terms of the annual kWh achieved on an isolated detached house with 50° pitched roof at the optimum site in Wick.

Table 19 shows the total 60-year cost for each turbine, including renewals and maintenance, and the annual equivalent cost as both actual and discounted costs. Table 20 shows the same data but taking into account the savings made by electricity generation or sales on the most preferable terms. As the net cost per annum for any option is still a cost, as opposed to a saving, we must conclude that, in cost terms alone, the value of electricity produced does not outweigh the cost of supplying and maintaining/replacing any of the turbines, even in the most favourable location considered in this study. (i.e. with the current costs there is no financial payback within the expected life of the systems).

Table 20: LCC data as Table 19 but allowing credit for electricity generated.

	60-year costs (zero discount)		60-year net present value (3.5% discount)	
	Total cost	Annual cost	Total cost	Annual cost
Micro-wind system 1	£2,973	£50	£2,942	£49
Micro-wind system 2	£2,742	£46	£2,354	£39
Micro-wind system 3	£1,679	£28	£3,363	£56
Micro-wind system 2 (no maintenance)	£5,850	£98	£3,829	£64

7   DISCUSSION AND CONCLUSIONS

Estimates have been made of the environmental impact of the manufacture, installation and maintenance of three typical micro-wind turbines suitable for mounting on domestic buildings taking account of some different installation and maintenance regimes.

A wind tunnel study, which is reported separately [3], was undertaken to investigate the effects on the wind speed of the house on which the turbine is mounted and of the presence of surrounding buildings. In principle, there are a huge number of interrelated variables in the building geometry and orientation, the turbine mounting positions, and the positions, types and orientations of adjacent buildings. For the purposes of this study, typical detached and terraced houses with 25° and 50° dual-pitched roofs were considered either standing alone or with other buildings 10, 20 or 50 m away (i.e. typical domestic houses). Airflow measurements were made for a number of different mounting positions and heights above the roof of the buildings studied.

A method for estimating the wind resource in urban environments has been developed using the BREVe software, which is normally used for calculating wind loads on buildings and is based on BS 6399 [11]. This work has produced scaling factors that have been applied to wind speed data for three different types of urban landscape in the UK. Manchester was chosen to represent a large inland city and Portsmouth and Wick as examples of different sized conurbations and coastal regimes.

The data from these studies have been used to estimate the expected annual electricity generation for some 'typical installations' of the three micro-wind turbines in the studied locations. These results have then been compared with the environmental impact of the manufacture, installation and maintenance of the turbines to calculate the CO_2 payback periods.

These studies have shown a large variation in the expected CO_2 payback periods from a few months in good locations to situations where they never pay back, in poor locations.

The wind resource in Manchester was found to be the lowest of the locations studied and relatively consistent across the city. The estimated CO_2 payback results for Manchester are the least favourable, with most situations studied giving no payback within the expected life of the systems. This is particularly the case for the most likely mounting conditions (i.e. with the turbine mounted up to 1.5 m above the roof of buildings which have similar buildings adjacent to them).

However, for the smaller cities (Portsmouth and Wick) the results show a wide difference between the best and worst locations, with the worst location being the city centre in both cases. In the best locations all the results indicate that the CO_2 payback is within the expected life of the systems. For the worst locations (city centres) the results are mixed and indicate that the ability to pay back the CO_2 is very dependent on the type of building and the proximity of other buildings.

A life cycle cost analysis of the systems included in this study suggests that, even in the most favourable location considered in the study, there is no financial payback within the expected life of the systems, with the current system and electricity costs.

In this study the three turbines were selected as being typical devices that may be used in the types of installation studied. However, this study is not intended to compare the performance of the three turbines and all three turbines were modelled in each situation, using the same assumptions about maintenance and expected service life, to give generic information, and without attempting to match the turbine to the actual conditions. In practice, turbine characteristics, such as the 'cut-in' and 'cut-out' wind speeds, may make one turbine more suitable for a particular application than another. For example, in locations like Manchester the ability of the turbine to produce power at low wind speeds is very important but the ability to generate at high wind speeds may be less important: however, the situation may be quite different for a windy location such as Wick.

It should be noted that the calculations do not take account of turbulence effects, and although the general effect of the town terrain and surrounding buildings is accounted for, it is assumed that the building on which the turbine is mounted is not overshadowed by larger buildings, trees or other obstructions close by. It is well recognised that high levels of turbulence will reduce the electrical output of the turbine and is likely to increase the requirements for maintenance. In the wind tunnel study, Blackmore [3] identifies that mounting turbines on buildings increases the turbulence and it is therefore considered that the predictions for the electricity generation and hence the payback periods given in this study are likely to be optimistic.

The study also highlights the fact that the maintenance regime required for the turbines has a very significant effect on the costs and environmental impact of the systems over their service lives. This is particularly the case for the types of building mounted turbines considered because of the difficulty of obtaining access to the turbine once it is mounted high above the roof of the building. Although the turbines are designed to have low maintenance requirements, they do have moving parts that will be in operation for a large proportion of their service life. They will also be subject to large and rapidly changing loads during heavy winds, which are likely to be exacerbated by the turbulence experienced close to buildings. It is therefore considered that a maintenance inspection at least every three years is a reasonable requirement to maximise the service life of the unit and to allow the necessary safety inspection of the blades, mountings, bearings etc. Much of the cost and environmental impact of this maintenance is due to the transport of the means of access (scaffolding or cherry picker) and the service engineer. These could be reduced by the provision of mountings that allow the turbine to be lowered to a safe level for maintenance. Also, if these turbines become more popular and are fitted in clusters, for example, to all of the buildings on a development site, then the cost and impact of both installation and maintenance could be considerably reduced as an engineer could install or maintain a number of turbines on a single visit.

It should also be noted that this study has only investigated micro-wind turbines on some typical domestic houses in urban environments. It has not looked at the mounting of turbines on tall buildings, which may give very different results for both the electricity generated, and the costs and difficulties associated with access for installation and maintenance.

This study has not investigated the structural aspects of mounting wind turbines on domestic buildings or the possible nuisance from noise or vibration that may be caused. However, it is considered that structural, noise and vibration considerations may also be significant factors in determining the practicality of using building mounted wind turbines.

Additionally, this work has highlighted the fact that the NOABL database appears to overestimate the wind resource for any terrain type other than open flat country with cut grass and that this overestimation is likely to be much more significant for built-up areas.

8   FURTHER WORK

This work has shown simple calculations using different manufacturer's power curves. Although the conditions under which these were obtained are, in most of the cases, uncertain, it is expected that the power curves were either obtained in wind tunnels or using free-standing mast-mounted turbines far away from obstacles and under low levels of turbulence.

The presence of the building on which the turbine is mounted has been shown to increase the level of turbulence that the turbine will experience. Due to the complexity of the subject, the dynamics of the turbine and its behaviour in higher than usual turbulence has been omitted from this study. It is recommended that further attention is paid to this issue.

In this study, the data for only three different Met stations have been considered. It is suggested that a similar analysis could be undertaken for other UK locations in order to extract conclusions representative of other locations.

It is suggested that a database that accounts for the UK terrain roughness could be developed (e.g. a modified version of the NOABL database). This could be done by combining the information provided by the BREVe software and its database of UK terrain roughness with the NOABL database to allow for a corrected wind speed database that accounts for the local topography and roughness of the UK terrain. This database would be more suitable for estimating the wind resource for small turbine installations for different terrain types, from farm lands to suburban and urban areas. It could also serve as a guideline for installers under the new UK Microgeneration Certification scheme and wind energy engineers in general.

Due to the scope of this project, it has only been possible to look at the electricity generated by three types of wind turbine located at a few points on a limited selection of building types in one orientation. Further work could include other types of turbine and a greater selection of locations and building types to give a systematic analysis of the influence that these factors would have in the electricity produced over the course of a year.

The configurations considered in this work are representative of low-rise buildings. The effects of surrounding buildings have been considered by including a fixed density building layout with similar height buildings. It is recommended that a similar exercise be undertaken for highly built-up areas and that medium to high-rise buildings are also considered. The tops of higher buildings will see higher wind speeds and it is thought that these may be more suitable locations for wind turbines, especially in highly built-up city centres such as London or Manchester.

It would be useful to validate the method presented here using monitoring data from real applications. Some DTI-sponsored field trials [1, 2] are currently underway, although the output from these studies are not expected until after autumn 2008. It would be possible to use data from smaller trials (e.g. the Warwick urban wind trial project [1]) as it becomes available. With validation, the method presented here could be shown to be a useful analysis tool, with the potential to considerably improve the accuracy of estimates of urban electricity generation.

9   REFERENCES

1. The Warwick Urban Wind Trial Project, managed by Encraft Ltd, 2007, www.warwickwindtrials.org.uk

2. **Energy Saving Trust.** Assessment of household wind turbines, www.energysavingtrust.org.uk/aboutest/news/pressreleasesarchive/index.cfm?mode= view&press_id=552

3. **Blackmore, P** (2007). *Micro-wind turbines: wind speeds over house roofs.* BRE Trust Report FB18, IHS BRE Press, January 2008.

4. **BRE** (2007) Methodology for environmental profiles of construction products. (In preparation).

5. **Rankine RK, Chick JP and Harrison GP,** (School of Engineering and Electronics, University of Edinburgh), Energy and carbon audit of a rooftop wind turbine, *Proc. IMechE* Vol.220 Part A: *J. Power and Energy,* pp643–54

6. **Swiss Centre for Life Cycle Inventories.** *The ecoinvent Database.* www.ecoinvent.ch

7. www.pre.nl/simapro

8. ISO 14044: 2006. Environmental management – Life cycle assessment – Principles and framework. International Standards Organisation, Geneva

9. http://cig.bre.co.uk/envprofiles

10. BREVe2, *Design wind speed software tool - an aid to the use of BS6399-2:1997.* Available from www.brebookshop.com.

11. BS 6399-2:1997 including Amendment 1:2002 and corrigendum 1:2002, *Loading for buildings. Part 2 Code of practice for wind loads,* British Standards Institution, London

12. **BERR** (2007), *The Department for Business, Enterprise & Regulatory Reform's wind speed database (NOABL database).* www.dti.gov.uk/energy/sources/renewables/ renewables-explained/wind-energy/page27708.html

13. **Burch SF, Makari M, Newton K, Ravenscroft F and Whittaker J** (2004). *Computer modelling of the UK wind energy resource: Phase II Application of the methodology,* ETSU WN7054, 1992

14. **Office of Government Commerce.** Procurement Guide 07, *Whole-life costing and cost management.* OGC, London, 2007. www.ogc.gov.uk/documents/CP0067AEGuide7.pdf.

15. ISO/DIS 15686-5 (2004) *Draft International Standard - Buildings and constructed assets - Service life planning - Part 5: Whole life costing.* International Standards Organisation, Geneva

16. www.wlcf.org.uk

17. www.bettergeneration.co.uk

APPENDIX A

UK MEAN ANNUAL WIND SPEED AT 25 m ABOVE GROUND LEVEL

From NOABL database. Copyright ETSU for the DTI 1999

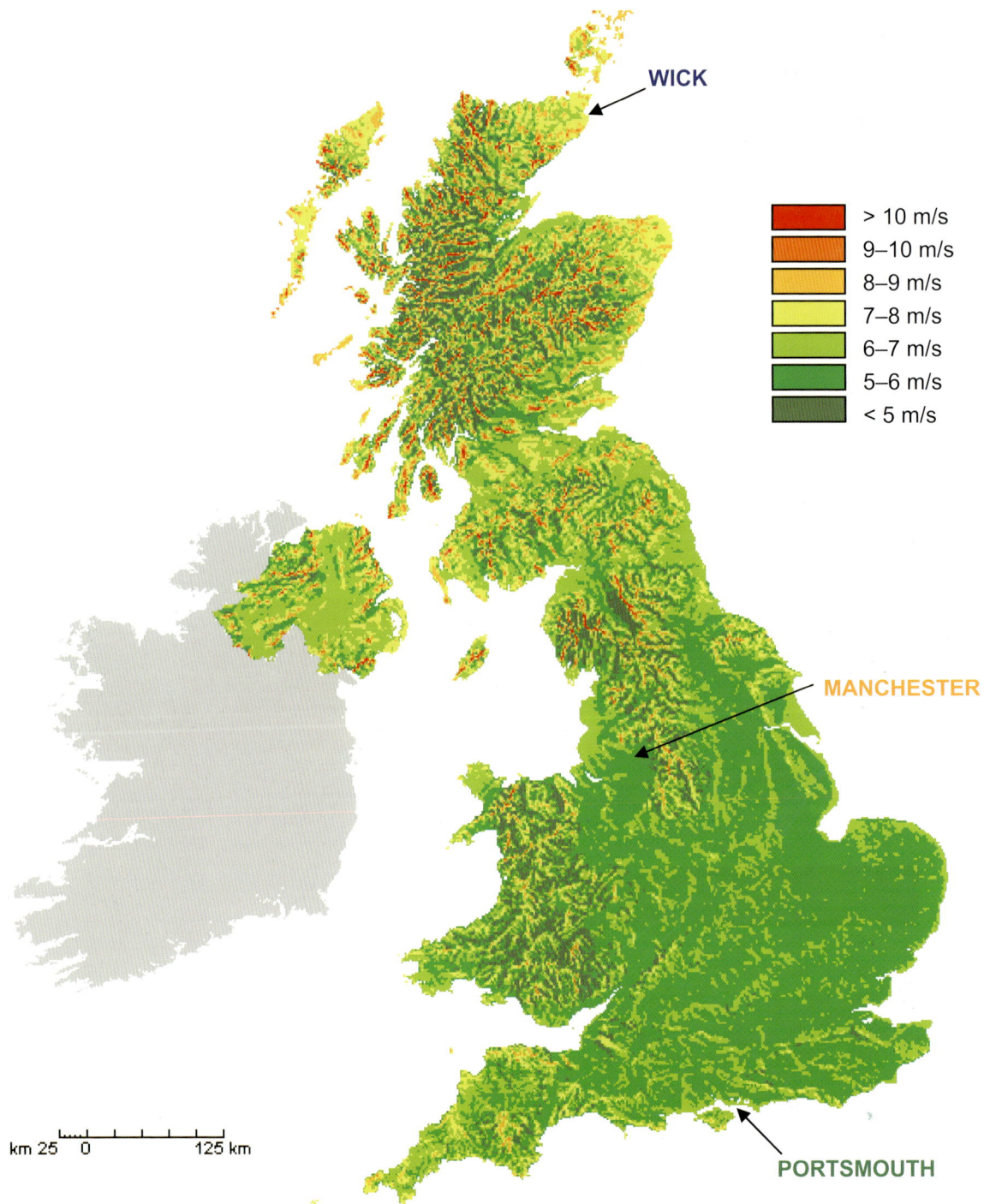

WICK

MANCHESTER

PORTSMOUTH

> 10 m/s
9–10 m/s
8–9 m/s
7–8 m/s
6–7 m/s
5–6 m/s
< 5 m/s

km 25   0   125 km

APPENDIX B
LOCATIONS FOR WHICH THE BREVe SCALING FACTORS WERE OBTAINED

Wind roses and city centres are also shown (see Figures 8–10).

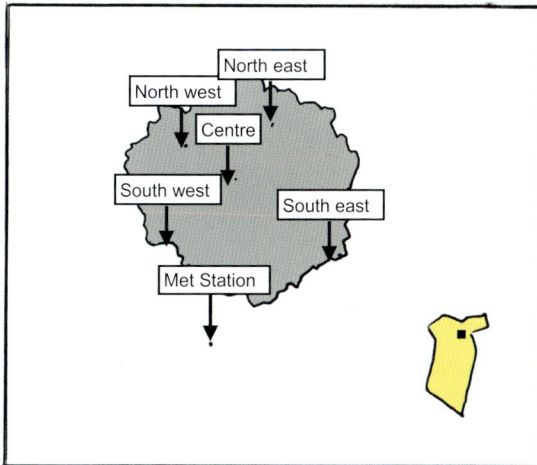

Manchester

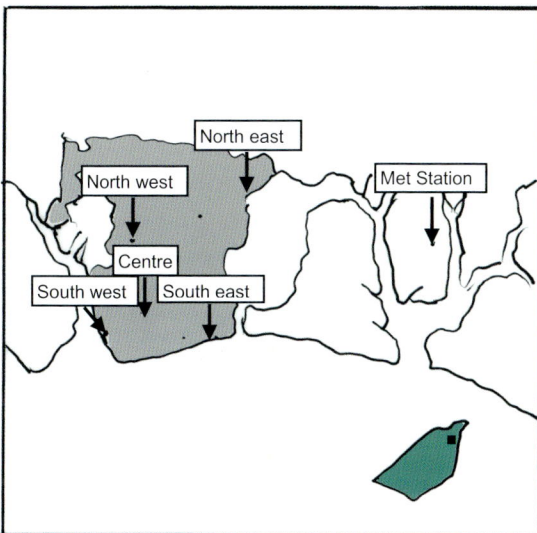

Portsmouth

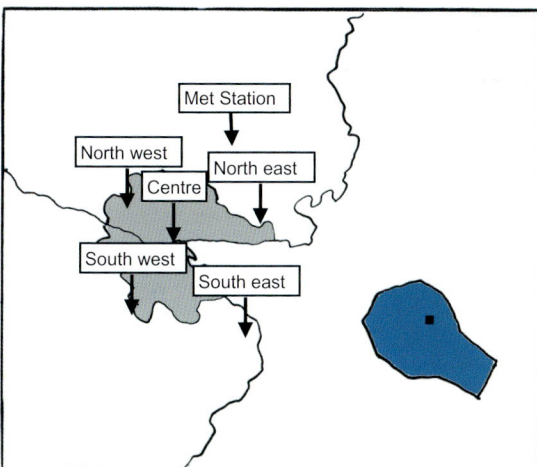

Wick

OTHER REPORTS FROM BRE TRUST

FB1 SUBSIDENCE DAMAGE TO DOMESTIC BUILDINGS
Lessons learned and questions remaining
R M C Driscoll and M S Crilly
September 2000

FB2 POTENTIAL IMPLICATIONS OF CLIMATE CHANGE IN THE BUILT
ENVIRONMENT
Hilary M Graves and Mark C Phillipson
December 2000

FB3 BEHAVIOUR OF CONCRETE REPAIR PATCHES UNDER PROPPED
AND UNPROPPED CONDITIONS
Critical review of current knowledge and practices
T D G Canisius and N Waleed
March 2000

FB4 CONSTRUCTION SITE SECURITY AND SAFETY
The forgotten costs
Bob Knights, Tim Pascoe and Alice Henchley
December 2002

FB5 NEW FIRE DESIGN METHOD FOR STEEL FRAMES WITH
COMPOSITE FLOOR SLABS
Colin Bailey
January 2003

FB6 LESSONS FROM UK PFI AND REAL ESTATE PARTNERSHIPS
Drivers, barriers and critical success factors
Tim Dixon, Alan Jordan, Andrew Marston,
James Pinder and Gaye Pottinger
November 2003

FB7 AN AUDIT OF UK SOCIAL HOUSING INNOVATION
Keith Ross, James Honour and Fran Novak
February 2004

FB8 EFFECTIVE USE OF FIBRE REINFORCED POLYMER MATERIALS IN
CONSTRUCTION
S M Halliwell and T Reynolds
March 2004

FB9 SUMMERTIME SOLAR PERFORMANCE OF WINDOWS WITH
SHADING DEVICES
Paul Littlefair
February 2005

FB10 PUTTING A PRICE ON SUSTAINABILITY
BRE Centre for Sustainable Construction and Cyril
Sweett
May 2005

FB11 MODERN METHODS OF HOUSE CONSTRUCTION
A surveyor's guide
Keith Ross
June 2005

FB12 CRIME OPPORTUNITY PROFILING OF STREETS (COPS)
A quick crime analysis – rapid implementation approach
Joan Oxley, Petra Reijnhoudt, Paul van Soomeren,
Calvin Beckford, Armando Jongejan and
Joachim Jager
November 2005

FB13 SUBSIDENCE DAMAGE TO DOMESTIC BUILDINGS
A guide to good technical practice
Richard Driscoll and Hilary Skinner
June 2007

FB14 SUSTAINABLE REFURBISHMENT OF VICTORIAN HOUSING
Guidance, assessment methods and case studies
Tim Yates
September 2006

FB15 PUTTING A PRICE ON SUSTAINABLE SCHOOLS
Anna Surgenor and Ian Butterss
Spring 2008

FB16 KNOCK IT DOWN OR DO IT UP?
Sustainable housebuilding: new build and refurbishment in
the Sustainable Communities Plan
Frances Plimmer, Gaye Pottinger, Sarah Harris,
Michael Waters and Yasmin Pocock
Spring 2008

FB17 MICRO-WIND TURBINES IN URBAN ENVIRONMENTS
An assessment
Richard Phillips, Paul Blackmore, Jane Anderson,
Michael Clift, Antonio Aguiló-Rullán and Steve Pester
November 2007

FB18 MICRO-WIND TURBINES
Wind speeds over house roofs
Paul Blackmore
Spring 2008

The reports can be ordered from IHS BRE Press, Willoughby Road, Bracknell, Berks RG12 8DW
Tel: 01344 328038, brepress@ihs.com or at www.BREpress.com